MUNICIPAL SOLID WASTES

PROBLEMS and SOLUTIONS

Edited by
**Robert E. Landreth
Paul A. Rebers**

CRC Press
Taylor & Francis Group
Boca Raton London New York

CRC Press is an imprint of the
Taylor & Francis Group, an **informa** business

CRC Press
Taylor & Francis Group
6000 Broken Sound Parkway NW, Suite 300
Boca Raton, Fl. 33487-2742

© 1997 by Taylor & Francis Group, LLC
CRC Press is an imprint of Taylor & Francis Group, an Informa business

First issued in paperback 2019

No claim to original U.S. Government works

ISBN-13: 978-0-367-44840-0 (pbk)
ISBN-13: 978-1-56670-215-7 (hbk)

Visit the Taylor & Francis Web site at
http://www.taylorandfrancis.com

and the CRC Press Web site at
http://www.crcpress.com

Library of Congress Cataloging-in-Publication Data

Municipal solid wastes: problems and solutions / edited by Robert E.
  Landreth and Paul A. Rebers.
        p.   cm.
    Includes bibliographical references and index.
    ISBN 1-56670-215-1
    1. Refuse and refuse disposal–United States–Congresses.
2. Refuse and refuse disposal–Congresses. I. Landreth, Robert E.
II. Rebers, Paul A.
TD788.M864   1996
    363.72′85′0973—dc20   96-31106
                                                                                  CIP

Library of Congress Card Number 96-31106

Typesetter: Pamela Morrell
Cover design: Dawn Boyd

Preface

Environmentalists of the '90s have been faced with the challenge of managing an ever-increasing solid waste issue. People are moving at an increasing pace, requiring new methods of technology transfer, mode of transportation, and living conditions. At the same time, this drive requires that certain packaging is required to prevent loss of product, and to improve durability. Elaborate separation tasks are usually not economically feasible.

These two ideas are not at conflict, but require innovative ideas and thinking in order to provide satisfactory answers to both sides of each issue. With this in mind, the American Chemical Society, Division of Environmental Chemistry presented a symposium on selected topics. Most of the papers presented are reported here. A few papers were added to complete the thought process. Each paper was current at the time of presentation and was updated by the chapter author to reflect rapidly changing conditions in environmental issues.

The book discusses an integrated approach to managing solid waste...the traditional reduce, recycle, combustion, and disposal approach. It should be recognized that not all of these steps will see use in a given community. Chapters 1 and 2 are concerned with glass and plastic recycling. The basic fundamentals are explained as well as the concern for quality and the issue of transportation.

Disposal of household hazardous waste and the legal implications are discussed in Chapters 3 and 4. Household hazardous wastes are defined, and the issues, the legal requirements, and solutions to the problems are all discussed. Chapter 5 discusses composting, a form of recycling, as well as feedstock, techniques, and legislation. An important aspect of composting is what to do with the final product. Special attention is given to the physical and chemical quality of the end product so as to market the product.

Chapters 6, 7, and 8 deal with scrap tire issues and methods utilized to minimize their disposal. While incineration plays a big part in scrap tire disposal, other civil engineering solutions are being developed and marketed. Increased use of these solutions will help in solving this difficult environmental problem.

Chapters 9 through 12 are concerned with combustion. The state of the art is discussed in Chapter 9, by identifying the volumes, regulations, waste-to-energy issues, and health and safety issues. Generally speaking, a sizable volume is required in a community to minimize the cost associated with a waste-to-energy facility. Chapter 10 discusses the role of a small town municipal solid waste incinerator. A comparison of operating cost provides some interesting results for the smaller communities. Chapter 11 discusses mercury and dioxin issues, a detailed discussion of sources for the two contaminants and their potential human exposure, an interesting set of data that deserves review. Chapter 12 discusses combustion ash, defines combustion ash and residuals, their physical and chemical characteristics, leachability characteristics, and treatment options that may be available, disposal options and potential utilization.

Chapters 13, 14, and 15 discuss landfilling of municipal solid wastes. Although not a popular option with a majority of people, it is nonetheless an essential means of economically disposing of wastes. Chapter 13 discusses all technical aspects of construction quality control/quality assurance for liners, focusing on compacted soil liners and geomembrane liners. Chapter 14 discusses cover systems for both Subtitle C and D and CERCLA landfills, the state requirements, alternative cover designs and special issues such as gas generation, migration, and control strategies. Chapter 15, the last, discusses design of Subtitle C and Subtitle D landfill containment systems.

The editors wish to thank the U.S. EPA Research Facility at Center Hill in Cincinnati, the engineering staff at Iowa State University and of the City of Ames for providing the encouragement to present this symposium. As a point of history, the first successful solid-waste-to-energy plant was constructed in Ames in 1975 because of the perceived need to have an improved method of disposal of solid waste.

The editors also want to personally thank each of the contributors and the many people who made this book a reality. We especially want to thank the American Chemical Society for allowing us to present this important symposium.

Robert E. Landreth
Paul A. Rebers

The Editors

Robert E. Landreth has recently retired from the U.S. Environmental Protection Agency where he worked for the last 33 years. During this time at the Agency, he was responsible for the formulation, planning, administration, and evaluation of a comprehensive national research program for proper treatment and disposal of municipal solid and hazardous waste on land. In this capacity he provided national and international expertise and technical advice to the Agency's operating program, regional offices, and foreign countries. He served as a focal point to executive, legislative, and legal inquiries to identify relationships between research and standard development documents, guidance documents, and land resource documents. He also provided technical assistance to private and international consultants on waste management design and operation.

Mr. Landreth is a member of ASTM D-35, the North American Geosynthetics Society, the Solid Waste Association of North America, and the Industrial Fabrics Association International. He also participated as a team member touring the nation presenting training courses explaining the Subtitle C and D regulations on land disposal. He was written many papers on land disposal research and has been responsible for a variety of state-of-the-art research reports.

Paul Rebers worked as an immunochemist from 1961 until his retirement in 1988, at the National Animal Disease Center in Ames, Iowa on fowl cholera, an important disease of chickens and turkeys. His work on the chemistry of the antigenic structure of the causative agent, *Pasteurella multocida*, contributed to an improved serologic test for typing the organism. This test was adapted as a standard diagnostic procedure by the U.S. Department of Agriculture for use in epidemiological studies of the organism. He worked with Dr. Michael Heidelberger for six years at Rutgers University and crystallized the repeating unit of the Type VI pneumococcal polysaccharide.

Dr. Rebers earned a B.S. and M.S. in Chemical Engineering, and a Ph.D. at the University of Minnesota. He has published over 60 papers on structure of polysaccharides and on the immunochemistry of animal disease. He was co-editor with Dr. C. Richard Cothern (U.S. Environmental Protection Agency) of the book, *Radon, Radium, and Uranium in Drinking Water* (published, 1990). Dr. Rebers was a member of the graduate faculty of Iowa State University.

Dr. Rebers belongs to the American Chemical Society, the American Society of Microbiology, Sigma XI, Phi Lamda Upsilon, and is an honorary member of Phi Zeta. He is listed in the 1995 American Association for the Advancement of Science (AAAS) Resource Directory of Scientists and Engineers with Disabilities; and served as a member of the City Council in Nevada, Iowa from 1990 to 1993.

Contributors

S. Keith Adams, Ph.D.
Department of Industrial and
 Manufacturing Systems
 Engineering
Iowa State University
Ames, Iowa

Bonnie Blaylock, M.A.
Center for Risk Management
Oak Ridge National Laboratory
Oak Ridge, Tennessee

Michael H. Blumenthal, M.B.A.
Scrap Tire Management Council
Washington, D.C.

Marc A. R. Cone, B.S., P.E.
State of Maine Department of
 Environmental Protection
Augusta, Maine

David E. Daniel, Ph.D.
Department of Civil Engineering
University of Texas at Austin
Austin, Texas

R Del Delumyea, Ph.D.
Millar Wilson Laboratory
Jacksonville University
Jacksonville, Florida

**John C. Even, Jr., Ph.D.,
P.E.**
Department of Industrial and
 Manufacturing Systems
 Engineering
Iowa State University
Ames, Iowa

Robert W. Hartley, B.S., P.E.
State of Maine Department of
 Environmental Protection
Augusta, Maine

Lynnann Hitchens, B.S.
Office of Research and Development
U.S. Environmental Protection
 Agency
Cincinnati, Ohio

Richard M. Kashmanian, Ph.D.
Office of Planning, Policy and
 Evaluation
U.S. Environmental Protection
 Agency
Washington, D.C.

Jonathan V. L. Kiser, M.B.A.
Environmental Consultant
Round Hill, Virginia

Robert M. Koerner, Ph.D.
Geosynthetic Research Institute
Drexel University
Philadelphia, Pennsylvania

Robert E. Landreth, M.S., P.E.
Municipal Solid Waste & RMB
U.S. Environmental Protection
 Agency
West Chester, Ohio

Karen Morrison, B.S.
State of Maine Department of
 Environmental Protection
Augusta, Maine

Paul Rebers, Ph.D.
Collaborator, Assoc. Professor, Ret.
Iowa State University
and U.S.D.A., A.R.S., Ret.
Ames, Iowa

Natalie U. Roy, B.A.
National Pollution Prevention Round
 Table
Washington, D.C.

John R. Serumgard, L.L.M.
Scrap Tire Management Council
Washington, D.C.

Joan Sulzberg, Ph.D.
Environmental Law Institute
Washington, D.C.

Curtis C. Travis, Ph.D.
Center for Risk Management
Oak Ridge National Laboratory
Oak Ridge, Tennessee

Edward C. Weatherhead, B.S.
Scrap Tire Management Council
Washington, D.C.

Carlton C. Wiles, B.S.
National Renewable Energy
 Laboratory
Golden, Colorado

Richard K. White, Ph.D.
Environmental Systems Engineering
Clemson University
Clemson, South Carolina

Table of Contents

*Dedicated by the authors to our grandchildren
and to future generations*

CHAPTER 1

Recycling Realities and the Glass Container: New Technologies and Trends

Natalie U. Roy

Soda–lime–silica glass — which is used to manufacture glass containers — is made from a few naturally occurring, readily available raw materials: sand, soda ash, limestone, dolomite limestone, and feldspar. They must meet stringent specifications with respect to grain size and purity. The glass manufacturing process is a relatively simple one. The batch, which is comprised of the required weights of each raw material, is thoroughly mixed and fed into the furnace melting chamber by one of several different types of charger systems.

The process energy resulting from the combustion of the fuel, and from electrical boosting, gradually raises the temperature of the glass batch. At times, up to 10% of the total energy added to the glass is in the form of electric heating that results in a reduction in the fossil fuel requirements. As the temperature increases, a variety of complex physical and chemical reactions take place. Initially, there is disassociation of some of the raw materials. Subsequent reactions between them results in the formation of some relatively fluid liquid phases. The latter react with each other and continue to dissolve the sand grains, which are the most difficult material to take into solution. During the process, the batch has been heated to approximately 2700°–2800°F in order to obtain the desired final quality. The "molten" glass is then molded into new glass packaging.

GLASS CONTAINER RECYCLING

In the early 80s, the glass container industry recognized that a solid waste crisis was imminent and aggressively began promoting recycling curbside collection programs. At this time, the industry had started to suffer serious market share loss, due in part to forced deposit laws in several states. Curbside collection of recyclables was viewed as a more comprehensive alternative that enhanced, rather than reduced, the marketability of the environmentally friendly glass container. Glass containers also lend themselves well to community recycling programs; they are 100% recyclable and can be recycled over and over again. Moreover, the recycling process creates no additional waste or by-products.

From a manufacturing standpoint, using cullet (used glass) can save wear and tear on furnaces, resulting in saving on maintenance. Cullet can be melted at a temperature lower than that necessary to melt virgin materials. This reduced melting point also allows glass container manufacturers to reduce energy input to furnaces (as much as $3 to $8 per ton, depending on utility rates and furnace size). Additionally, glass recycling has environmental benefits, such as reduced air emissions and also prolongs furnace life. The only material glass container manufacturers use more than cullet is sand.

In additional to being extremely recyclable, glass has other environmental benefits. These include the following:

- A simple production process, which results in less pollution than competitive packaging materials;
- Few hazardous inputs, which means less toxic emissions;
- No serious chemical accidents resulting from the production process; and
- Consumption of domestically plentiful non-toxic raw materials.[1]

In 1993, glass containers, which comprise 4% of the municipal solid waste stream by weight, were recycled at a rate of 35% in the U.S. This means that more than one-third of the glass containers available for consumption in this country were recycled back into glass containers and other valuable items such as fiberglass, or returned as refillable bottles.

From 1987 through 1993, domestic glass container manufacturing plants increased their purchases of cullet by over 95% — totaling over 2,400,000 tons in 1992. In contrast, the number of glass plants has declined by 40% in the last 12 years. This loss in plants is attributed to consolidation of the industry and more efficient machinery, as well as changes in packaging uses.

A decrease in the number of plants directly impacts recycling transportation costs. However, even with fewer plants available to receive cullet, the industry was still able to dramatically increase the amount of recycled glass it purchased. Currently there are 70 glass container manufacturing plants operating in the U.S.

Though recycling of glass has skyrocketed, the sudden growth of curbside collection and drop-off programs (5,500 at last count and growing) has created a glut of used materials for manufacturers of all commodities. One of the biggest recycling issues facing the glass container industry is quality control — ensuring that incoming cullet meets industry specifications. As a result, the glass container industry is investing significant resources into improving glass processing technologies.

Quality Concerns

Recyclables are basically industrial raw materials. All industrial raw materials must meet quality specifications. The glass container industry cannot use amber or green glass to make clear containers. Glass used for recycling must be color-sorted and contaminant free. Unfortunately, some communities and some system vendors choose to promote systems that allow for unacceptable levels of contamination. The Glass Packaging Institute (GPI) believes the best way to collect all recyclables is through curbside collection and sorting. Curbside separation of recyclables by the collector will deliver to all user industries the greatest amounts of high quality recyclables. The industry is investing significant resources into improving glass processing and recycling technologies. These innovations include automated color separation, mechanized ceramic sorting, organic color coating, and cullet pulverization technologies. The industry is also experimenting with using more green cullet in amber glass batches to increase recycled glass usage.

While these technological advances are being explored by the industry, there are presently no economically viable mechanical systems that effectively and efficiently color-separate glass or pull ceramics or other non-metallic contaminants out of delivered loads. Organic color coating is also in the experimental stage. Some companies are experimenting with color coating, the most familiar project being the Brandt Process which has been publicized extensively. Color coating would eliminate color separation by staining flint bottles with organic colors that would then burn off in the furnaces. The process though has still not been fully developed and has not met the glass industry's current production line needs.

In some cases, glass container plants have installed state of the art glass processing equipment (beneficiation units), which remove metals and other contaminants. These units cost anywhere from $500,000 to $1,000,000 each. The glass processing business is expanding with more independent third party processors entering the industry.

There are also other major technological innovations taking place in the glass processing business. For example, the New York State Energy Research and Development Authority, with funding and technical assistance support from several private entities including the GPI, unveiled a new glass beneficiation system that is able to optically sort out ceramics and separate non-ferrous metals from the cullet that has been collected for recycling.

The non-ferrous metal separator in this experimental processing system removes aluminum, lead, and small contaminants, including the rings from bottlenecks. Following this process, the cullet is then moved to the optical ceramic sorter which removes unwanted ceramics by releasing quick jets of air activated by an infrared light that detects the ceramic's opacity. There is also research taking place investigating automated means to separate container glass by color, which would make residential color separation unnecessary. These are just some of the efforts going on across the country to develop improved methods of processing furnace-ready cullet.

Transportation Issues

Transportation plays an important role because it can make or break a community recycling effort. For example, in some parts of the country there are no glass plants, resulting in prohibitively high costs for transporting the materials across long distances. This also causes shortages of recycled materials for some of our glass plants. To remedy the situation, the industry is working with communities to explore ways of making transporting cullet less costly, such as backhauling or shipping by rail. Depending on market locations, some recyclers backhaul different materials, transporting recyclables from one destination and hauling back other materials on the return trip.

Glass container plants across the country also have different color specifications. Some plants only produce flint containers, while others may produce amber and flint. To alleviate this problem, some companies have started a swapping program whereby they ship by rail the cullet they do not need in exchange for a color needed at their plant. Transportation concerns have spawned new applications for cullet in secondary materials. Secondary markets that are locally based are important for some communities that cannot afford to ship their cullet a long distance.

Secondary markets include fiberglass, glassphalt, roadbed, reflective beads, decorative glass, and drainage. In 1992 approximately 450,000 tons of container cullet were used in the production of these materials. One of the most common and visible secondary markets is glassphalt, an asphalt mix that uses crushed glass as an ingredient developed as a method of using commingled cullet. Fiberglass, another common secondary market for cullet, is predominantly used in the form of glass wool for thermal and acoustical insulation.

Lightweighting

Most so-called solid waste management hierarchies make source reduction their top priority. Source reduction is generally defined as reducing the quantity or concentration of objectionable components of solid waste before the solid waste is generated. Examples of source reduction usually include minimizing, down-sizing or lightweighting packaging and eliminating certain objectionable materials as package constituents.

Through advances in design and manufacturing technology, the glass container industry is constantly reducing the weight of glass containers. For instance, according to a 1992 Franklin and Associates study entitled, "Analysis of Trends in Municipal Solid Waste Generation," from the period of 1972 to 1987, on average non-returnable glass containers were reduced in weight by 44%.[2]

While source reduction is an admirable goal, it has limited application to solid waste management. Packages cannot be perpetually lightweighted. Sooner or later, all packages will reach their optimum weight. In addition, source reduction by weight will not address diminishing landfill capacity. After all, landfills fill up by volume, not by weight. Even more importantly, replacing recyclable materials with non-recyclable materials in the name of source reduction is not progress and will have an adverse environmental impact. The glass container, with its easy recyclability and its low volume impact on landfills, is an ideal container in terms of source reduction.

Comprehensive Recycling

Recycling is not the total answer to the solid waste problem. However, efficiently operated recycling programs can easily divert 35% or more of municipal solid waste away from disposal. GPI believes that recycling programs should be comprehensive and convenient. GPI believes that curbside collection systems offer the opportunity to collect the greatest amount of recyclables in the most cost efficient manner. However, it is imperative that communities and recyclers operate programs that are glass-friendly and that result in color separated, contaminant-free material.

The State of New Jersey demonstrates enlightened solid waste management through its Mandatory Recycling Act. New Jersey law requires counties and municipalities to recycle 60% of the municipal solid waste stream by 1995. While the law does not require curbside collection systems, this is by far the most popular recycling collection method in New Jersey. The success of the New Jersey approach can be measured by the fact that within three years of the adoption of the Mandatory Recycling Act of 1987, New Jersey reached a 46% recycling rate (in 1990 — the most recent statistics available) and leads the nation in recycling. The success of comprehensive approaches for glass recycling in New Jersey is shown by its 68% recycling rate for glass. New Jersey, along with other comprehensive recycling states such as Washington and Rhode Island, is successful because its law targets *all* recyclables.

Non-Comprehensive Approaches

The glass container manufacturing industry strongly opposes mandatory beverage container deposit systems (also known as forced deposits or bottle bills). These systems place a deposit on a small percentage of the waste stream (beverage containers are only 3.5% of municipal solid waste by weight and 2.5% by volume). Ironically, deposit advocates often argue that deposits will solve *all* of America's solid waste problems. Forced deposit systems result in higher beverage prices because of the increased costs to manufacturers, distributors, and retailers for labeling, handling, storage, accounting, reporting, transporting, and other costs.

Deposit systems also have an adverse impact on the environment. In bottle bill states, beverage markets shift to less recyclable materials due to material handling problems caused by deposit systems. Why does this happen? The retailer wants to reduce costs to its lowest possible level. As a result, the retailer demands that bottlers provide more beverages in lighter packaging that can be crushed or stored without breaking, thereby taking up less room in the store. The irony is that the 100% recyclable glass container is replaced with packaging such as plastics, which is less recyclable and not as environmentally friendly.

Recycling programs also lose valuable revenue from beverage containers in forced deposit states. It is normal for beverage containers to constitute 19% of the materials collected in a comprehensive recycling program and supply over 73% of the revenue for the program. Deposits negatively affect the economic viability of recycling. Citizens also become confused about recycling as they have to remember which materials go back to the store and which are recycled at curbside. Recycling suffers. It is no surprise that when recycling rates in comprehensive recycling states are compared with recycling rates in adjacent forced deposit states, the comprehensive states always have higher recycling rates. For example, New Jersey is higher than New York, and Rhode Island is higher than Massachusetts and Connecticut.

Utilization Rates and Recycled Content

The glass container industry, through its network of regional recycling managers across the country, is committed to increasing both cullet recovery and post-consumer content. However, with more state and federal legislation on the horizon, *it is critical that the glass industry have flexibility in order to achieve mandated targets*. Glass should be able to count cullet tonnage used in secondary glass applications towards recycled content or recovery rates.

In California, for example, glass is required to meet strict post-consumer content requirements. Alternatively, the plastics industry is allowed to pick from a menu of less rigid options that permits the counting of secondary applications. To maintain a level playing field and continue the upward trend in glass recycling, the industry needs to be allowed to count the growing glass secondary market towards recycling mandates, just as the plastics industry is allowed to do. Allowing credit for secondary market cullet also means the glass container industry would not be penalized when recyclers generate poor quality material, that only has lower valued uses, such as road aggregate or glassphalt.

Recovery and recycled content requirements must take into account that cullet meet quality requirements and the fact that other industries, in addition to the glass container industry, use cullet as a raw material. It is also critical that there is adequate available supply of cullet that meets industrial quality requirements. Recycling requirements must be based on real world technological limitations. The U.S. is a net importer of glass containers. Fairness requires that all glass containers sold in the U.S. be subject to similar requirements. As such, recycling requirements will have an impact on international trade and might be difficult to enforce against glass container manufacturing plants in other countries.

Finally, recycling requirements should apply equally to all competing packages. Mandating higher recycling requirements for some packages places the packages with the higher requirements at a potential competitive disadvantage. Industries with the higher requirements are required to spend more resources creating an infrastructure for collecting and remanufacturing their packages.

Volume-Based Fees

Most Americans have no idea what it costs to dispose of their garbage. Few Americans actually pay for solid waste management services based on the amount of solid waste they generate. This trend goes completely counter to the way that most Americans pay for electrical and water services. When people are presented with higher bills for water consumption or energy consumption, they normally conserve their use of water and energy. Surely the same will happen with solid waste collection services.

In the few American towns that bill residents for solid waste services based on the amount of garbage collected for disposal, recycling programs are thriving and the amount of solid waste being disposed of is diminishing. Volume-based fee systems may be a variable can rate such as in Seattle, or a bag or tag system such as is found in Perkasie, PA, High Bridge, NJ, and other communities. A bag system works, for example, by charging residents for each bag of garbage thrown away. Volume-based systems, instead of being an unusual method of charging for solid waste services, should become the normal method of charging for solid waste services.

Solid Waste Management Economics

One of the greatest barriers to increased recycling lies in misleading and inaccurate accounting systems for solid waste management costs. In spite of our attempts to aid communities in determining the costs of recycling, the true costs of other solid waste management systems are usually unknown. In most cases, landfill tipping fees do not cover closure, post-closure, corrective action, and replacement costs. As a result, inaccurate comparisons of the cost of recycling with the cost of landfilling are often made to the detriment of recycling. It is imperative that landfill tipping fees show their full true costs. When this occurs, efficiently operated recycling programs will prove to be the least expensive method of solid waste management for the recyclable portion of the waste stream.

Green Glass Recycling

Green glass is an important part of the packaging mix for both product preservation and marketing. Colored glass protects certain food and beverages from degrading in sunlight and artificial light. The darker colors help extend the shelf-life of certain products, which is a vital characteristic of any package. Green glass is also an important marketing tool used to identify certain products for consumers.

Annually, U.S. glass container manufacturers produce 1,289,050 tons of green glass, which represents 12% of all glass containers produced in the U.S. Out of all of the glass manufacturing facilities in the U.S., 16 plants are the primary green glass producers.[3]

A 1993 GPI report produced for the State of New Jersey showed that although the market for green cullet has been volatile, the current market is sound. There are no known piles of uncontaminated green cullet which glass recyclers cannot market. However, not every community is collecting green glass in its recycling program, and it is possible there may be market difficulties in the future. Green glass recycling markets face problems not faced by amber or flint glass markets. These factors include transportation costs, plant locations, the amount of imported products packaged in green glass, and seasonal demand for green containers.

In the New Jersey report, GPI recommends five practical solutions to improving green glass recycling. These recommendations include encouraging states and other jurisdictions to create cullet storage areas, enforcing cullet quality standards and expanding secondary markets, and can be easily and cost-effectively implemented. To obtain a copy of the green glass report, contact the GPI office at (202) 887-4850.

Environmentally Sound Green Glass

Ironically, for all its alleged market difficulties, green glass is the most recyclable of the three major colors. Larger quantities of flint and amber cullet can be utilized when manufacturing green glass than is possible when manufacturing the other two colors (see ASTM glass color specifications chart attached to report). Because of its ability to utilize more amber and flint cullet, green glass manufacturing generally posts the highest recycling rates. Nationwide emerald green glass (the most common color of green) has an average recycling rate of 45%, compared to 26% for amber and 18% for flint. The rates for amber and flint are also reflective of their stricter color specifications.

There are other green shades produced, in much smaller amounts than emerald. Other colors include dead leaf, georgia, champagne, and spring green. These colors are primarily used in wine bottles. Since there is a limited amount of available separated cullet in these specialty green shades, recycling rates are much lower than for emerald. Blue bottles are also produced for the growing new age beverage market. Just as with the specialty green shades, the amounts are still minimal when compared to the other colors. Many plants request that residents put their blue containers in with their green, but some companies with different specifications may request other sorting arrangements.

One other process worth noting is called fritting. Some plants, to gain the flexibility to meet customers' demands for different green shades, especially for limited runs, will color flint glass in the forehearth. This is done after the molten glass has exited the melter prior to the container forming process. This is not an exceedingly common practice and those furnaces where fritting takes place were not included in the green plant totals since they do not provide a market for green glass cullet.

For years, the glass container industry has worked to develop technology to increase the use of three-color mixed cullet in production to minimize sorting. Years ago the spotlight was on eco-glass, a mix of all three colors. The glass industry can produce such a bottle, but to date there has been little commercial or consumer interest.

REFERENCES

1. Cole, H. S. and Brown, K. A., *Advantage Glass,* Glass Packaging Institute, Washington, D.C., 1993.
2. Franklin and Associates, Analysis of Trend, in *Municipal Solid Waste Generation, 1972 to 1987,* Final Report, January 1992.
3. Roy, N. and Preede, K., *Green Glass Recycling,* Glass Packaging Institute, Washington, D.C., 1993.

CHAPTER 2

Post-Consumer Plastics Recycling

S. Keith Adams and John C. Even

INTRODUCTION

Plastics in the Total Waste Stream

Plastics typically constitute between 4 and 15% of post-consumer and munic-ipal solid waste on the basis of weight, but often require up to 30% of the volume needed for landfilling because of their low density and widespread use in the manufacture of containers for many consumer products.[3,8-10]

The U.S. generates approximately 180 million tons of municipal solid waste each year, or approximately 450,000 tons each day. A conservative estimate of the percentage plastic constituent of this total equal to 6.5% would suggest a generation rate of 11.7 million tons or 23.4 billion pounds of plastic waste per year in the U.S. Data[3,8] shows a national percentage estimate by the U.S. Environmental Protection Agency of 6.5% and an estimate derived by the authors for the state of Iowa of 9.3%. The percentage of plastics in the municipal waste stream has been growing steadily since 1958 to 57 billion pounds in 1988, at an annual rate of 10.3%[5] and is likely to continue to grow.

In a study conducted by the authors for the City of Ames, IA during 1990 to 1991, a nine-month analysis of the incoming waste stream for the city's Resource Recovery Plant showed the percentage of plastics to be 7.2%, up from 4.2% derived in a similar study at the same plant in 1977.[2] Based on the national estimate given earlier, a 7.2% fraction would suggest a national plastic waste generation rate of 26 billion pounds per year, somewhat below the EPA estimate. The U.S. EPA has estimated that approximately 29 billion pounds of plastic are disposed of in the municipal solid waste stream each year and that only 1.1% of the plastic waste stream, or 400 million pounds, are recovered.[10] Table 2.1 provides a breakdown of sources and quantities for selected plastic products manu-factured and recovered by weight, according to U.S. EPA estimates. Only 1.1%

Table 2.1   Generation, Recovery, and Disposal of Plastic Products, 1988 [U.S. EPA, 1990]

Product category	Generation (10⁶ tons)	Recovery (10⁶ tons)	(%)	MSW discards (10⁶ tons)
Durable goods	4.1	<0.1	1.5	4.1
Nondurable goods				
Plastic plates and cups	0.4	—	—	0.4
Clothing and footwear	0.2	—	—	0.2
Disposable diapers	0.3	—	—	0.3
Other misc. nondurables	3.8	—	—	3.8
Subtotal	4.6	—	—	4.6
Containers and packaging				
Soft drink bottles	0.4	0.1	21.0	0.3
Milk bottles	0.4	—	<0.1	0.4
Other containers	1.7	—	—	1.7
Bags and sacks	0.8	—	—	0.8
Wraps	1.1	—	—	1.1
Other plastic packaging	1.2	—	—	1.2
Subtotal	5.6	0.1	1.6	5.5
Total plastics	14.4	0.2	1.1	14.3

From Hegberg, B. A., et al., *Mixed Plastics Recycling Technology*, Noyes Data Corporation, Park Ridge, NJ, 1992. With permission.

of these products were recovered or recycled, indicating a need to encourage plastics recycling through product design, collection, preparation quality assurance and transportation programs, remanufacturing, and finally consumer acceptance and use of recycled plastic products.

A similar study conducted for the Plastics Recycling Foundation yielded an estimate of 340 million pounds of plastic recycled in 1989.[11] This total includes consumer products not accounted for in the EPA study, but still represents only 2.5% of domestic packaging used in 1989 for the six major thermoplastics used and less than 1% of all virgin thermoplastic resins. Only one specific type of plastic container, the PET beverage bottle, commonly collected in curbside programs and repurchased through legislated deposit policies has achieved a sizeable recycling rate of 23% in 1988 and 29% (175 million pounds in 1989).[12] Similar programs need to be developed for HDE and other plastics.

Types, Sources, and Uses of Post-Consumer Plastics

Post-consumer plastics are typically classified into seven standard numerically designated categories for purposes of recycling, based on resin type. Six basic types are designated; the seventh denotes a mixture. The resin type for a given plastic refers to its polymer chemical composition and structure. The seven polymer classifications are listed in Table 2.2 along with their common applications in consumer products. Each type of plastic resin has its own recycling technology and mark.

Table 2.2 Common Symbols Used to Designate Recycled Plastic Types

Society of plastics industry symbols	Plastic resin types	Common applications
⚠1	Polyethylene terephthalate	Beverage containers, boil-in-pouches, processed meat packages
⚠2	High-density polyethylene	Milk and water bottles, detergent bottles, toys
⚠3	Poly Vinyl chloride	Food wraps, surgical gloves, piping blister packaging
⚠4	Low-Density polyethylene	Shrink wrap film, bag films, garment bags
⚠5	Polypropylene	Margarine and yogurt containers, caps for containers, medicine bottles, car seats
⚠6	Polystyrene	Disposable plastic silverware, egg cartons, fast food packaging, video cassettes, televisions
⚠7	Other	Multi-resin containers

The Society of Plastics Industry and other representative organizations have compiled data given in Table 2.3 showing the annual demand for each type of resin and its percentage relative to total demand which includes co-polymers or mixed plastics (category 7). Co-polymers present additional problems and should be avoided as container material wherever possible. Categories 4, 3, 2, and 5 account for 57.6% of the total and include a large portion of consumer packaging, excluding PET (category 1) widely used for soft drinks, yet represent only 3.6%. Part of the problem in recycling beverage and food containers is caused by the U.S. Food and Drug Administration's restriction against their re-use as beverage and food containers following recycling. This eliminates a major prime recycling market (same type of product) for these materials and requires the use of secondary recycling for products with lower quality standards such as plastic lumber.

Polyurethane and phenolic resin plastics are not designated by an SPI code number. These plastics are used primarily in industrial products such as circuit boards, electrical insulation, and various protective and insulation coatings.

As stated previously, only a small percentage of post-consumer plastic waste is recycled nationwide in the U.S. However, local and regional data vary over a wide range, and overall the percentage is growing steadily. Utility Plastic Recycling, Inc., located in Brooklyn and serving the New York City area has eliminated some of the major economic barriers to plastics recycling by combining the manufacturing of products from recycled plastics with a material collection and recovery facility at one location.[14] New York City has required residential recycling since 1989. Public education, strong local markets such as this, and legal enforcement have provided a reliable high quality feedstock for the plant. Utility Plastic Recycling is a member of a group of companies, Waste Management of New York (WMNY). Other members include: JLJ Recycling, Commercial Recy-

Table 2.3 Major Resin Markets, with Emphasis on Packaging, 1989 [Modern Plastics, 1990]

Market category	Quantity
Appliances	1,197
Building	11,390
Electrical/Electronics	2,202
Furniture	1,190
Housewares	1,362
Packaging	
Closures	
HDPE	81
LDPE	32
PP	420
PS	190
PVC	36
Other	20
Subtotal	779
Coatings	
HDPE	51
LDPE	730
PET	10
PP	30
PVC	21
Other	251
Subtotal	1,093
Containers	
HDPE	3,400
LDPE	311
PET	1,049
PP	454
PS	1,306
PVC	352
Other	146
Subtotal	7,018
Film	
HDPE	541
LDPE	3,421
PP	612
PS	211
PVC	310
Other	107
Subtotal	5,202
Packaging (total)	14,092
Toys	729
Transportaion	2,200
Major market total	34,362

Note: The abbreviations used in this table are spelled out in Table 2.2.

From Hegberg, B. A., et al., *Mixed Plastics Recycling Technology,* Noyes Data Corporation, Park Ridge, NJ, 1992. With permission.

cling Technology, and Evergreen Recycling. WMNY operates a 325,000 square foot complex occupying two city blocks of abandoned warehouses in Brooklyn. Their 60 ton per hour capacity process has created more than 30 jobs. It provides industrial feedstock in the form of baled HDPE and also produces traffic barricades and detour cones from recycled PVC. The physical and economic aspects of this process are discussed later in this chapter.

A mandatory beverage container recycling law or "bottle bill" has a very strong effect on the composition of the municipal solid waste stream. This is evidenced in the difference in waste stream composition between communities located in the same geographic region with similar population demographics and standards of living, differing in the matter of having or not having mandatory beverage container recycling. One example of this occurs when comparing data for Duxbury, MA, a community of 14,000 located 30 miles south of Boston, with data for Rhode Island. Massachusets requires beverage container recycling while Rhode Island does not. As a result of this difference, plastics constitute 9.2% of Rhode Island's waste stream but less than 0.1% of Duxbury's.[15] Based on the Rhode Island data, PET containers make up 0.5% of the waste stream but Duxbury recycles only clear milk jugs, a portion of the total PET content, recycling of other plastics has not been found to be cost effective. This points to another problem, the economic feasibility of collection transporting and recycling in small communities and rural areas, as discussed later.

There are large markets for plastic resin in the packaging industry which represent great potential for recycled plastic flake, provided that the overall recycling process can compete on a quality and total cost basis (including transportation). Table 2.3 lists resin quantities according to various end product uses with a detailed breakdown of plastic types used in packaging by category of use and annual demand for each resin type within each use.[13] Containers and wrapping film dominate packaging applications, accounting for 12.2 of the 14.1 billion pounds of plastic used annually for packaging in the U.S. within the demand for containers and film, HDPE, LDPE, PET, and PS account for three fourths of the annual demand. These plastics need to be separated and recycled through voluntary drop-off and curbside collection programs.

Tables 2.4 and 2.5 show total plastic disposal in the U.S. and sources of plastic disposed of in the municipal solid waste (MSW) stream in 1988, according to a study by Franklin Associates.[16] In Table 2.4, differentiation is made between plastics disposed of primarily through MSW and those disposed of through other routes. Large quantities of HDPE, LDPE, PP, and PS are available for recycling from the MSW stream. These four plastics account for 21.2 of the 23.2 billion pounds of plastic in the MSW stream, or 91.3%. As shown in Table 2.3, there is high demand for HDPE, LDPE, and PS for use in containers. Polypropylene (PP) is a frequently disposed plastic, as seen in Table 2.4, indicating a quantity larger than its total resin demand listed in Table 2.3. Among plastics disposed of primarily through non-MSW routes, PUR and PVC are the major constituents, totaling 3.1 billion of 3.6 billion pounds, or 85.7%. The 1.8 billion pounds of non-MSW annually disposed PVC is greater than the 719 million pounds of PVC resin, shown as annual demand in Table 2.3. As shown in Table 2.4, 23.2 billion

Table 2.4 U.S. Plastic Disposal, 1988 [Franklin Associates, 1990]

Resin	Total disposed (10⁶ lbs)	Non-MSW disposal (10⁶ lbs)	%	MSW disposal (10⁶ lbs)	%
Plastics disposed primarily through MSW routes					
ABS	1,093.3	383.1	35.0	710.2	65.0
HDPE	6,528.8	975.3	14.9	5,553.5	85.1
LDPE	7,690.8	577.2	7.5	7,113.6	92.1
PET & PBT	1,475.5	176.2	11.9	1,299.3	88.1
PP	5,274.0	1,016.9	19.3	4,257.1	80.7
PS	4,767.9	529.7	11.1	4,238.2	88.9
Subtotal	26,830.3	3,658.4	13.6	23,171.9	86.4
Plastics diposed primarily through non-MSW routes					
Acrylic	686.0	663.3	97.7	22.7	3.3
Nylon	461.6	329.2	71.3	132.4	28.7
Phenolic	2,975.1	2,869.0	96.4	106.1	3.6
PUR	2,794.8	1,510.7	54.1	1,284.1	45.9
PVC	7,566.0	5,799.4	76.7	1,766.6	23.3
Unsat. Polyester	1,319.3	1,183.0	89.7	136.3	10.3
Urea & melamine	1,459.2	1,346.7	92.3	112.5	7.7
Subtotal	17,262.0	13,701.3	79.4	3,560.7	20.6
Total	44,092.3	17,359.7	39.4	26,732.6	60.6

From Hegberg, B. A., et al., *Mixed Plastics Recycling Technology,* Noyes Data
Corporation, Park Ridge, NJ, 1992. With permission.

pounds (46%) of plastic were disposed of in MSW, compared with 26.7 billion
pounds through non-MSW routes (54%).

Table 2.5 lists the quantities and percentages of types of plastic in the municipal waste stream derived from residential, commercial, and institutional sources.
Similar overall profiles of plastic types are shown for all three sources. As in
Table 2.4, LDPE, HDPE, PP, and PS dominate in terms of quantity and account
for 80.4, 79.3, and 74.1%, respectively, of residential, commercial, and institutional MSW plastic. Resource recovery processes should be designed to recover
these dominant types of plastic. Curbside recycling for these plastics in separate
bins should also be encouraged. Mixed plastic containers such as often seen in
1 and 2 liter soft drink bottles composed of a PET upper portion and polypropylene or other base portion should be discouraged. Caps typically made of polypropylene or polystyrene should be removed and recycled separately.

When plastics are disposed of in a general MSW stream, they constitute a
small portion by weight (typically 5–9%). Separating them from other materials
and differentially isolating different types of plastic can be prohibitively expensive. Table 2.6 lists data[6] for the city of Milwaukee, showing the overall composition of the MSW stream with plastics constituting 8.4% of the total, but no
specific resin type comprising more than 0.5% by weight. It is very difficult to
achieve reliable separation of specific plastic constituents with an acceptable level
of quality (lack of contamination by other materials or other plastics). Hand
sorting of plastics is very unlikely to be cost effective.

Table 2.5 Sources of Plastic Disposed in MSW in the U.S., 1988 [Franklin Associates, 1990]

Resin	Quantity disposed (10⁶ lbs)	Residential (10⁶ lbs)	%	Commercial (10⁶ lbs)	%	Institutional (10⁶ lbs)	%
LDPE	7,113.6	4,606.5	64.8	1,621.8	23.1	865.3	12.2
HDPE	5,553.5	3,783.3	68.1	1,007.1	18.1	763.1	13.7
PP	4,257.1	2,138.6	50.2	1,439.9	33.8	678.6	15.9
PS	4,238.2	2,308.0	54.5	1,202.5	28.4	727.5	17.2
PVC	1,766.6	1,033.2	58.5	347.6	19.7	385.7	21.8
PET & PBT	1,299.3	664.9	51.2	286.0	22.0	348.3	26.8
PUR	1,284.1	877.1	68.3	260.9	20.3	146.1	11.4
ABS	710.2	229.0	32.2	370.0	52.1	111.1	15.6
Unsat. polyester	136.2	68.2	50.0	40.9	30.0	27.3	20.0
Nylon	132.4	98.4	74.3	23.6	17.8	10.5	7.9
Urea and melamine	112.5	88.3	78.5	12.5	11.1	11.7	10.4
Phenolic	106.1	63.7	60.0	32.3	30.4	10.1	9.5
Acrylic	22.7	9.1	40.1	5.7	25.1	8.0	35.2
Total	26,732.6	15,968.3	59.7	6,670.8	25.0	4,093.3	15.3

From Hegberg, B. A., et al., *Mixed Plastics Recycling Technology*, Noyes Data Corporation, Park Ridge, NJ, 1992. With permission.

Table 2.6   Residential Waste characterization for Milwaukee with Emphasis on Plastics [Englebart, 1990]

MSW Component	Composition of waste stream (weight %)
Dirt, diapers, and fines	6.6
Ferrous metals	5.1
Food	16.0
Hazardous materials	0.3
Glass	8.3
Leather	0.1
Multi-material packages	0.4
Non-ferrous metals	1.6
Paper	31.0
Plastics	
PET containers ≥ 1 liter	0.4
PET containers ≤ 1 liter	0.1
HDPE colored, tubs	0.5
HDPE , milk, water, juice bottles	0.5
Rigid plastics	1.9
Flexible bags and films	4.5
Polysyrene	0.4
Other (durable goods, non-pkg.)	0.1
Subtotal	8.4
Rubber	0.7
Rubble	0.8
Textiles	4.6
Yard waste	
Leaves	0.0
Grass clippings	13.8
Brush/branches/weeds	1.5
Wood (lumber,strumps, pallets)	0.8
Total	100

From Hegberg, B. A., et al., *Mixed Plastics Recycling Technology,* Noyes Data Corporation, Park Ridge, NJ, 1992. With permission.

Many individual communities have conducted publicly funded recycling studies that have included plastic recycling and an analysis of recovered plastics by type and quantity based on curbside collection or separation from the general MSW stream.[6] It is important that each community conduct its own studies and not rely solely on national or regional data since local factors have a strong influence on the waste stream composition.

For example, a study of plastics in the MSW stream in Hamilton, Ohio yielded an estimate of 6.2% plastic by weight for 1990. The profile of plastics by type from this study is given in Table 2.7. The national average portion of plastic in the MSW stream was estimated at 9.2% by the U.S. EPA[10] for the same year. The category listed as "other" in Table 2.7 included such items as HDPE containers for laundry detergent, bleach, margarine and butter, and motor oil, and also LDPE containers such as bottles for shampoo and PVC clear bottles as well as bottle caps made of polystyrene.[17] The city of Chicago conducted a four season study of its waste stream during 1989 and 1990 and estimated the amount of

Table 2.7 Plastic in MSW of Hamilton, Ohio
           [Peritz, 1990]

Plastic type	% by Weight in MSW	% by Weight of plastics
PS Foam	0.5	8
PET Soda bottles	1.0	16
HDPE milk/water bottles	0.7	11
Bags (LDPE film)	1.8	29
Other[a]	2.2	36
Total	6.2	100

[a] Plastics contributed that were not specified as part of the
collection program.

From Hegberg, B. A., et al., *Mixed Plastics Recycling Technology*, Noyes Data Corporation, Park Ridge, NJ, 1992. With permission.

plastic at 9.4% by weight. Of this amount, 0.8% was PET beverage containers, 1.0% was HDPE containers and 7.6% comprised all other types.[18] Higher percentages of plastic are likely from specialized waste streams such as exist in airports and theme parks. A study conducted at the Miami, Florida International Airport, based on a solid waste stream of 35 tons per day produced an estimate of 12.5% plastic by weight. The plastic profile for this waste stream is given in Table 2.8. A large portion of the airport waste material was recyclable. Paper, corrugated cardboard, and glass comprised 45.4% of the waste stream; virtually all of this is marketable. High yield large quantity recycling at heavily used public and tourist facilities is feasible and economical because of the local concentration of people, material sources, and types and organized, pre-planned waste collection, including the possibility of pre-sorting by the customer to a limited extent.

Table 2.8 Composition of Miami, Florida
           Airport Plastic Waste Stream
           [Peritz, 1990]

Product	Weight (%)
Film plastic	40.6
Clear cups	16.5
Foams	10.6
Translucent cups	6.0
PET beverage bottles	4.6
Rigid (white) PS	4.4
Utensils	4.4
Rigid coffee cups	2.1
Mixed (other) plastic	3.3
High impact cups (dairy)	0.8
Straws	0.6
Non-plastic residue	6.1
Total	100

From Hegberg, B. A., et al., *Mixed Plastics Recycling Technology*, Noyes Data Corporation, Park Ridge, NJ, 1992. With permission.

Table 2.9 Composition of Collected Plastic Beverage Bottles [Morrow and Merriam, 1989]

Bottle type	South Plainfield, NJ (weight %)	Mount Olive, NJ (weight %)
Population in collection study	20,000	22,000
Program type	drop-off	curbside
Number of samples	10	11
Time period of collection	4/88–1/89	5/88–2/89
PET beverage bottles	51.37	55.87
HDPE milk/water bottles	30.10	33.23
Other non-beverage bottles	18.53	10.90
Composition of "Other non-beverage bottles"[,a]		
HDPE detergent bottles	14.74	7.09
HDPE motor oil bottles	0.63	1.01
PVC bottles	0.75	1.05
PET 2 liter	0.49	0
Other plastic	1.34	1.56
Unknown plastic	0.58	0.19

[a] Plastics contributed that were not specified as part of the collection program.

From Hegberg, B. A., et al., *Mixed Plastics Recycling Technology,* Noyes Data Corporation, Park Ridge, NJ, 1992. With permission.

Less information is available on the collection of mixed plastics. Most of this collected at curbside is comingled with other materials and must be separated at a materials recovery facility (MRF). Instructions to the public regarding the deposit of categories of plastic or plastic containers are very important. In a study conducted by the Center for Plastic Recycling Research (CPRR) at Rutgers University, curbside and drop-off mixed plastics were analyzed.[19] In this study, the only items specified for collection were beverage bottles. The composition consisted of 50% soda pop bottles, 30% milk and water bottles and 10–20% unspecified plastic by weight. The composition of samples taken in two communities: Mount Olive and South Plainfield, NJ, are shown in Table 2.9. Curbside collection was used in Mount Olive while drop-off collection was used in South Plainfield. Significant differences between the two communities are shown in the composition of "Other Non-Beverage Bottles" or items not specified as collection items. The high percentage of non-specified plastics using the drop-off method shows a need for public education and instructions that will be clearly understood. In the same study, two other communities used handout sheets that specified "beverage containers only" which conflicted with instructions printed on 20 gallon recycling bins labels "all plastic bottles." This caused an increase in non-specified plastics up to 20–25% by weight, nearly twice the 10.9% obtained in the Mount Olive study.

The average quantity by weight and volume of PET and HDPE bottles set out per household per week for single family homes in suburban communities in New Jersey in 1988 was estimated at 0.45 lb and 0.30 lb, respectively.[20] The respective weekly weights and volumes of various materials per household are

Table 2.10   Compostion of Recyclables Collected Weekly per Household in New
             Jersey [Rankin et al, 1988]

Recycle stream	Weight (lbs/setout)	Weight (%)	Density (lbs/yd³)	Volume (gallons/setout)	Volume (%)
Newspaper	7.75	48.4	500	3.1	23.7
Glass bottles	6.0	37.5	700	1.7	13.0
Metal cans	1.0	6.3	144	1.4	10.7
Aluminum cans	0.5	3.1	49	2.1	16.0
Plastic bottles, uncrushed					
PET (60%)	0.45	2.8	40	2.3	17.6
HDPE (40%)	0.30	1.9	24	2.5	19.1
Total per setout	16.0	100%		13.1	100%

Total MSW generated[a] (lb/household/week)   = 63.5
MSW recycled {(5%)   = 16.0/63.5 = 25.2

[a] This is based on 1000 lbs of MSW generated/person/year and 3.3 people per household.

From Hegberg, B. A., et al., *Mixed Plastics Recycling Technology*, Noyes Data Corporation, Park Ridge, NJ, 1992. With permission.

shown in Table 2.10. These figures can serve as a guide for estimating required truck volume and weight capacities for curbside collection. In this study, comingled collection of plastics, ferrous metal, aluminum, and glass was recommended for a single container with another container devoted to newspapers. The metals, plastics, and glass require mechanical separation at a MRF.

A study of curbside collection was conducted by the Council for Solid Waste Solutions and by local government agencies for the Minneapolis–St. Paul area in 1990.[21] In this case, plastics recycling was added to existing recycling programs. Several types of collection vehicles were used in addition to several types of instructions to residents. Details of the study are discussed in Chapters 2 and 3 of Hegberg et al.[6] On one set of routes (A) only plastic soft drink and milk bottles were collected. On a second set of routes (B), all plastic bottles were collected, and on the third set (C), all types of rigid plastic containers were collected. The A, B, and C routes were operated by the local public works department. In addition, another route operating in Minnetonka collected plastic milk, water, soft drink, and detergent bottles (except bleach). This route was operated by Waste Management, Inc., a private corporation. Compositions and quantities of plastics collected on these routes plus another covering three suburbs called the Hennepin County Recycling Group (HRG) are shown in Table 2.11. Non-plastic contaminants are not listed. These data are believed to be applicable to many medium and large cities throughout the midwestern U.S.

Other municipal studies conducted in Walnut Creek, CA and Akron, OH are described in Chapter 2 of Hegberg et al.[6] The same chapter contains an extensive list of major types of plastic, estimated quantities generated per person per year, curbside participation rate, set-of rate for particular plastics, pounds collected, sources of information, and comments regarding each community study.

Table 2.11 Plastic Composition and Collection Amounts in Minneapolis Area Pilot Programs (pounds per household served per year) [CSWS, 1990]

Types of plastic	A	Minneapolis route		Minnetonka route	Hennepin route
Collection area Plastics specified	Milk and soft drink bottles	All plastic bottles (B)	All rigid bottles (C)	Milk, water, soft drink, and detergent bottles	Milk and soft drink bottles
Plastics collected					
PET bottles	2.21 (40%)	4.90 (27.7%)	4.43 (19.8%)	2.38 (23.4%)	3.25 (36.8%)
PET non-bottles	0.0	0.24 (1.4%)	0.28 (1.2%)	0.01 (0.1%)	0.0
HDPE clear bottles	3.01 (57.9%)	5.74 (32.5%)	5.46 (24.4%)	5.22 (55.8%)	5.44 (61.5%)
HDPE color bottles	0.0	2.62 (14.8%)	3.59 (16.1%)	1.26 (13.5%)	0.0
HDPE non-bottles	0.0	0.15 (0.8%)	0.17 (0.8%)	0.0	0.0
PVC	0.0	0.49 (2.8%)	0.50 (2.2%)	0.03 (0.4%)	0.01 (0.1%)
Composites	0.0	0.23 (1.3%)	0.22 (1.0%)	0.01 (0.1%)	0.0
Unknown bottles	0.0	0.34 (1.9%)	0.83 (3.7%)	0.04 (0.4%)	0.02 (0.3%)
PP	0.0	0.38 (2.2%)	0.23 (1.0%)	0.02 (0.2%)	0.0
PS	0.0	0.19 (1.1%)	0.58 (2.6%)	0.01 (0.1%)	0.01 (0.1%)
Lids and caps	0.03 (0.05%)	0.55 (3.1%)	1.04 (4.7%)	0.18 (1.9%)	0.05 (0.6%)
Unknown non-bottles	0.0	0.70 (4.0%)	2.48 (11.1%)	0.04 (0.4%)	0.01 (0.1%)
Plastic films	0.0	0.04 (0.3%)	0.01 (0.0%)	0.0	0.0
Non-plastics	0.03 (0.5%)	1.09 (6.2%)	2.54 (11.4%)	0.15 (1.6%)	0.05 (0.5%)
Total	5.2 (100%)	17.7 (100%)	22.4 (100%)	9.4 (100%)	8.8 (100%)
Plastics portion of recyclables collected (%)	3.8%	5.1%	6.5%	3.7%	3.8%

From Hegberg, B. A., et al., *Mixed Plastics Recycling Technology*, Noyes Data Corporation, Park Ridge, NJ, 1992. With permission.

Summary

Plastics typically account for 6 to 9% of the MSW stream by weight. They can constitute 4 to 14% by weight of recyclable material in a planned recycling program. If collected specifically, plastic film can comprise as much as 25 to 40% by weight of all plastics collected.[6] When all types of plastic bottles are collected, up to twice the amount of plastic by weight generally will be collected as will be collected when only PET and HDPE beverage bottles are included. It is advisable for any community interested in plastics recycling to conduct a study to estimate the MSW composition and the quantities and types of plastic that are recyclable. As a general figure, studies have indicated that a program to recycle all rigid plastics from residential areas will yield approximately 30 lbs per household per year. Recycling of plastic bottles will yield approximately 20 lbs per household per year. If only plastic beverage bottles are recycled, the annual yield will typically be approximately 15 lbs per household on a per household served basis. Nonspecified plastics may amount to as little as 1% by weight when only PET and clear HDPE bottles are collected, but as much as 10 to 20% by weight for a mixture of all plastic bottles or multicolored HDPE bottles. Non-plastic contaminants are generally between 1 and 10% by weight.

Many communities are conducting curbside collection programs which include plastics, particularly plastic beverage bottles. For example, as of August 1990, approximately 110 municipal districts had implemented curbside collection in Illinois, serving 600,000 households.[6] Of these, 43 municipal districts collect post-consumer plastic from 221,000 residences. Details of plastic recycling in Illinois are given in Chapter 3 of Hegberg et al.[6]

PLASTICS RECYCLING PROCESS TECHNOLOGY

Levels of Recycling

There are three basic levels of recycling for plastics denoted as primary, secondary and tertiary. *Primary recycling* is the processing of secondary plastic (post-consumer or post-industrial) into a product with physical properties and functional utility similar to those of the original plastic product. Primary recycling is suitable for thermoplastics since they do not have chemically bonded or inter-locking polymer chains and can be melted and remolded. *Secondary recycling* is the processing of post-consumer or post-industrial plastic into materials or products having qualities that are inferior to those of the original products. Secondary processing technology is very similar to that used to process virgin resins. Consistent, uniform composition and high levels of quality (low contamination levels) are necessary. Typically a homogeneous resin or a mixture containing recycled and virgin resins (virgin pellets and recycled plastic flake) is melted and fed into an injection molding or extrusion process. This method has been used primarily in making products from recycled PET and HDPE beverage bottles and also HDPE milk jugs.[5] Table 2.12 lists a number of products[23] made from recycled

Table 2.12 Estimated Markets for Recycled Plastic Resins (millions of pounds)

Polyethylene teraphthalate product applications	Market size		High-density polyethylene product applications	Market size	
	1987	1992		1987	1992
Fiber	90	180	Bottles		
Injection molding	25	160	(nonfood)	—	115
Extrusion	25	130	Drums	—	25
Non-food grade containers	—	30	Pails	20	65
Structural foam molding	—	30	Toys	—	15
Paints, polyols, other chemical sheet	10	20	Pipe	30	80
			Sheet	—	25
Stampable sheet	—	30	Crates, cases, pallets	—	105
Other	—	10	All other	4	130
Total — polyethylene teraphthalate	150	590	Total — high-density polyethylene	54	560

From the Center for Plastics Recycling Research, Rutgers, The State University of New Jersey, New Brunswick, 1987, as published by Curlee, T. R. and Das, S., *Plastic Wastes: Management, Control, Recycling and Disposal*, Noyes Data Corporation, Park Ridge, NJ, 1991. With permission.

PET and HDPE with market data for 1987 and 1992 (projected in 1987). Secondary recycling technology is well developed and cost-effective. Resulting products are marketable. More could be produced and sold if more recycled plastic were available. Homogeneous resins (uniform plastic types) are needed for this class of products.

When mixed plastic resins (several plastic types) are used in secondary recycling, the resulting products are of lower overall uniformity and quality in terms of chemical and physical variables. As is done in processing homogeneous resin plastics, mixed resin plastics are heated, usually by a combination of friction and pressure above the melting points of the predominant resin types and either extruded under heat and pressure or molded into desired product configurations. Plastic flakes or granules that do not melt in the process or other contaminants such as paper, rubber, or aluminum are captured and embedded in the plastic product. These serve as fillers and do not contribute to physical stress-resistant properties. Other additives such as more fillers, colorants, flame retardants, or chemical stabilizers are blended in to provide desired product characteristics and material quality. Typical products resulting from secondary mixed resin plastic recycling include:

- plastic lumber
- animal feeders
- car stops and speed bumps
- pallets
- dock plates (marine or loading dock)
- detour markers
- security barriers
- pipe guides
- noise baffles
- work tables

- park furniture
- flood control barriers
- cable reels
- man-hole covers
- drain gratings
- playground equipment

and many others.

Mixed resin secondary recycling processes are becoming more numerous in the U.S. They are an especially popular form of recycling in Japan and Germany where landfilling is not an option. There are engineering and economic challenges in this type of plastic recycling. These include: maintaining dependable and consistent quality, limited markets, and low-valued products in relation to production cost. Greater variety and quantity of marketable products are needed in addition to reductions in processing costs. Buyers must learn to value the durability of these products and pay prices commensurate with their true value and production costs.

Tertiary recycling involves the recovery of basic chemical ingredients or fuel oil from plastics. Pyrolysis or heating in the absence of oxygen is a common form of tertiary recycling. Volatile monomers are vaporized, yielding oils similar to naptha and char material composed of carbon, ash, and other residue. The nature and potential use of products derived from tertiary recycling depends upon the composition of plastic feedstock and the conditions under which pyrolysis is performed. Useful products include combustible gases as well as liquids suitable for use as fuels under approved conditions.[24] British Petroleum Chemicals (BPC) is building a pilot plant to produce oil that is very similar to naptha.[4] Many plastic resins are derived from the naptha fraction of crude oil, particularly from the ethylene and propylene structured chains. BPC has projected a yield of 80% from recycled plastic. An additional 10 to 15% will be burned as gas to provide heat for the endothermic process in which recycled plastic is pyrolyzed at 750 to 1100°F into vaporized paraffins that are condensed to form the oil. Feedstock to the reaction vessel consists of plastic flake approximately 3/4 to 1 inch in diameter.

Polystyrene decomposes into styrene and various aromatic compounds. Polyvinyl chloride (PVC) breaks down into a variety of hydrocarbons and also hydrogen chloride. The latter is removed separately. The reactor operates as a fluidized bed with hot combustion gases rising through and over sand which rises and falls in a boiling type of motion. Flakes of polymer coming in contact with this hot bed have their chemical chains broken apart thermally under a high rate of heat transfer from the sand to the polymer. Contaminant solids and filler materials such as titanium dioxide-based pigments are deposited in the sand. The BPC plant is expected to process over 200 lbs of plastics per hour. Given sufficient markets for the oil derived from plastics, this type of recycling may be more cost effective than traditional multiple resin secondary recycling, which always results in products with quality characteristics that are inferior to those of the original virgin plastics used as feedstock.

Types of Plastic, Properties, and Uses

Plastics are composed of polymers referred to as resins. They are synthetic materials derived from petroleum distillates or from natural gas. They begin as monomers or individual fluid molecules that are formed by chemical reactions into solid polymer molecules consisting of long repeating chains in two or three dimensions. When the chains do not bond, or interlock in a third dimension, the resulting plastic is known as a thermoplastic and can be remelted and reformed. When the chains are chemically bonded in three dimensions, the resulting plastic cannot be remelted and is known as a *thermoset plastic* or *thermoset*. Thermosets are generally hard, rigid, and insoluble, and infusible.[5] Because of the three-dimensional structure, thermosets are stronger than thermoplastics at high temperatures. Because of their inability to be remelted or fused, they cannot be reshaped once they are formed. Thermoset plastics include polyurethanes and phenolics.

Feedstock chemicals for producing high volume plastics in the U.S. along with derived polymer types are listed in Table 2.13 (a). Polymer types, their densities, softening or melting temperatures, properties and applications are listed in Table 2.14.[6] Compatibility of polymers in terms of acceptable commingling is charted by level according to paired combinations in Table 2.15.[25] Mechanical separation of plastics by type has not yet been perfected to the point where reliable separation producing acceptable purity levels is feasible. Hand sorting is often uneconomical and has been found to produce carpal tunnel syndrome and other repetitive motion-induced disorders. Table 2.14 suggests the separation of plastic flake by water flotation since some polymers have densities greater than water (>1.00 gm/cm^3) causing them to sink while those with lower densities are skimmed from the surface. Even if this is done, many incompatible combinations still remain as shown in Table 2.15, unless other separation methods are used prior to flotation.

Table 2.13   Feedstock Chemicals for the Production of High Volume Plastics [U.S. EPA, 1990a]

Feedstock chemical	Possible polymer product
Acetylene	PVC, PUR
Benzene	PS, PUR, ABS
Butadiene	PUR, ABS
Ethylene	HDPE, LDPE, PVC, PS, ABS, PET, PUT, polyesters
Methane	PET, PUR
Napthalene	PUR
Propylene	PP, PUR, polyester
Toluene	PUR, polyester
Xylene	PS, PET, ABS, polyester, PUR

From Hegberg, B. A., et al., *Mixed Plastics Recycling Technology,* Noyes Data Corporation, Park Ridge, NJ, 1992. With permission.

Table 2.14   Commodity Thermoplastic Characteristics

Resin type	Density (g/cm³)	Softening or melting range (°C)	Properties/applications
LDPE	0.910–0.925	102–112	Largest volume resin for packaging; moisture-proof, film transparent
LLDPE	0.918–0.942	102–112	Use generally grouped with LDPE
HDPE	0.940–0.960 0.950(colored bottles) 0.960 (clear bottles)	125–135	Tough, flexible, and translucent material, used primarily in packaging; product examples include milk and detergent bottles, heavy-duty films, wire and cable insulation
PP	0.90	160–165	Stiff, heat and chemical resistant, used in furniture and furnishings, packaging and others; product examples include drink straws, fish nets, butter tubs, auto fenders
PS	1.04–1.10	70–115	Brittle, clear, rigid, easy to process, used in packaging and consumer products; product examples include foamed take-out containers, insulation board, cassette and compact disc cases
PET	1.30–1.40	255–265	Tough, shatter and wear resistant, used primarily in packaging and consumer products; product examples include soft drink bottles, photographic and X-ray film, magnetic recording tape. Shipment strapping
PVC	1.30–1.35	150–200	Hard, brittle and difficult to process, but processed easily using additives: a wide variety of properties and manufacture techniques are possible using differing copolymers and additives; product examples include sheet bottles, house siding, cable insulation

From Hegberg, B. A., et al., *Mixed Plastics Recycling Technology,* Noyes Data Corporation, Park Ridge, NJ, 1992. With permission.

Process Technology

Following the separation of a given thermoplastic and its grinding into flakes, the plastic can then be remelted and reformed into new products. Three types of molding are used to reform the plastic. These are *extrusion molding*, *blow molding*, and *injection molding*. In the first type, extrusion (confined, directed compression) forces the plastic through a shaping die in a continuous operation as shown in Figure 2.1. The plastic feedstock may enter the extruder in a molten state but usually is solid upon entering and is melted and pressurized inside the extruder.[6] The incoming feed may be in the form of powder, flake, or other granulated state following grinding. The typical plastic extruder consists of a horizontal screw feeder that rotates inside a thick cylinder with the infeed hopper located at one end and the shaping die attached to the discharge area at the other

Table 2.15 Compatibility of Polymers [McMurrier, 1990]

Polymer type	LDPE	LLDPE	ULDPE/VLDPE	Ethylene copolymers	HDPE	PP	EPM/EPDM	PS (general purpose, high impact)	SAN	ABS	PVC	NYLON	PC	ACRYLIC	PBT	PET
LLDPE	1															
ULDPE/VLDPE	1	1														
Ethylene copolymers	1	1	1													
HDPE	1	1	1	1												
PP	4	2	[1]	2	4											
EPM/EPDM	4	4	[1]	3	4	1										
PS (general purpose, high impact)	4	4	4	4	4	4	4									
SAN	4	4	4	4	4	4	4	4								
ABS	4	4	4	4	4	4	4	4	1							
PVC	4	4	4	[2]	4	4	4	4	2	3						
NYLON	4	4	4	[1]	4	4	[1]	4	4	4	4					
PC	4	4	4	4	4	4	4	4	2	2	4	4				
ACRYLIC	4	4	4	[3]	4	4	4	4	4	4	4	4	4			
PBT	4	4	4	[2]	4	4	4	4	4	4	4	4	1	4		
PET	4	4	4	[3]	4	4	4	4	4	4	4	3	1	4	1	
SBS	4	4	4	4	4	4	4	1	3	2	3	3	4	4	4	4

Note: Compatibility designations: 1 = excellent, 2 = good, 3 = fair, 4 = without compatibility. Designations [1], [2], [3] = compatibility level depending on composition.

From Curlee, T. R. and Das, S., *Plastic Wastes: Management, Control, Recycling and Disposal*, Noyes Data Corporation, Park Ridge, NJ, 1991. With permission.

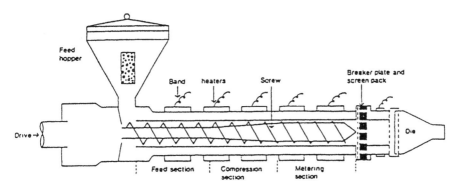

Figure 2.1 A single screw extruder [Morton-Jones, 1989]. The machine consists mainly of an Archimedian screw fitting closely in a cylindrical barrel with just enough clearance to allow its rotation. Solid polymer is fed in at one end and the profiled molten extrudate emerges from the other. Inside, the polymer melts and homogenizes. Twin screw extruders are used where superior mixing or conveying is important. (From Hegberg, B. A., et al., *Mixed Plastics Recycling Technology*, Noyes Data Corporation, Park Ridge, NJ, 1992. With permission.)

end. A combination of externally applied heat (when needed) and work done on the plastic by the shearing and compressive forces of the screw feeder melts the plastic and, at the same time, mixes it to create a uniform temperature before it is forced through the die. Profile extrusion is a variation of the general extrusion process. In this case, products having a continuous length are made. Examples include garden hose and plastic pipe.

Blow molding is the second general type of plastic forming process. There are several variations of this process including *extrusion blow molding*, *injection blow molding*, and *stretch blow molding*. In *extrusion blow molding*, the melted resin is extruded as a tube into the air. A blow pipe (or blow pin) is then inserted into the tube which itself is inserted into a mold cavity. Air pressure applied to the blow pipe expands the melted tube against the mold cavity and also cools the melt. The mold is then opened and the formed bottle is ejected and trimmed of excess plastic. Extrusion blow molding is generally used to produce plastic bottles with a capacity of more than 0.25 liters. Coextrusion makes it possible to operate multiple extruders simultaneously on a common die for processes that required layering or addition of colorants, where permeation across the bottle wall must be prevented.

Injection blow molding is a process wherein the molten plastic is injected into a cavity (parison cavity) around a core rod. The parison is then transferred to a blow mold cavity where the bottle is blown and cooled. This method is generally used in making bottles with a capacity less than 0.5 liters. Advantages of injection blow molding include a product free of scrap and suitability for manufacturing intricate shapes. It is unsuitable for making containers that have handles. Figure 2.2 illustrates a three-station injection blow molding machine.[26]

Stretch blow molding is a process wherein plastic is heated and pre-formed in a blow mold as in Figure 2.3. A center rod stretches the pre-formed plastic

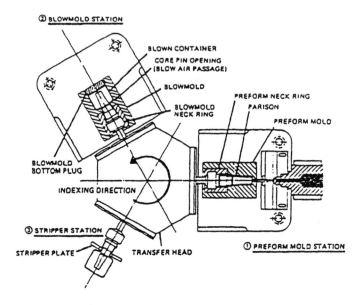

Figure 2.2   Three station injection blow modling machine [Miller, 1983]. The parison is injection molded on a core pin (instead of as a tube in free air, as with extrusion blow molding) at the preform mold station (1). The parison and neck finish of the container are formed there. The parison is then transferred on the core pin to the blow mold station (2) where air is introduced through the core pin to blow the parison into the shape of the blow mold. The blow container is then transferred to the stripper station (3) for removal. (From Hegberg, B. A., et al., *Mixed Plastics Recycling Technology,* Noyes Data Corporation, Park Ridge, NJ, 1992. With permission.)

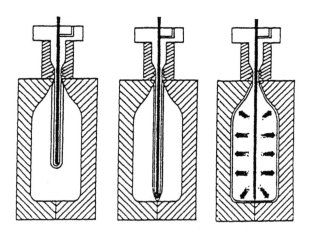

Figure 2.3   Stretch blow molding of bottles [Miller, 1983]. A heated preform melt is positioned in the blow mold. A center rod descends, stretching the preform with axial orientation. Blown air expands the preform into the mold, forming the bottle with radial orientation. (From Hegberg, B. A., et al., *Mixed Plastics Recycling Technology,* Noyes Data Corporation, Park Ridge, NJ, 1992. With permission.)

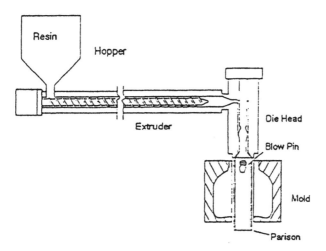

Figure 2.4 Extrusion blow molding of a bottle [Miller, 1983]. Hopper, reciprocating extruder
screw, die head, extruder parison, and molds shown in open position. The
extruder rotates and reciprocates continuously, providing continuous mixing of
resin with intermittent extrusion of parisons. (From Hegberg, B. A., et al., *Mixed
Plastics Recycling Technology,* Noyes Data Corporation, Park Ridge, NJ, 1992.
With permission.)

linearly (downward in Figure 2.3). Air is then blown into the mold, expanding
the plastic melt radially against the walls of the mold. Stretch blow molding
utilizes the crystal formation behavior of the resin and therefore requires pre-
conditioning of the plastic with respect to temperature. Stretching and cooling
must occur rapidly. PET soft drink bottles are formed using stretch blow molding.
Some PET bottles are also formed using extrusion blow molding.[6] The process
is illustrated in Figure 2.4.[26]

Injection molding is the third general type of molding used to form solid parts
at high rates of production. It also permits close tolerances and can be used to
produce very small parts difficult to produce by other methods. Scrap loss is also
very low in injection molding as compared with other molding processes since
scrap can be ground up and used again. In injection molding, molten plastic is
forced into the mold from an extrusion tube or barrel where it is held under
pressure. After the molded part has cooled and solidified, the mold is opened and
the part ejected.[6] Injection molding is illustrated in Figure 2.5.

Plastic Film Production

Plastic films are typically produced using a blown film extrusion or slit die
extrusion process. When blown film extrusion is used, a tubular film is produced.
Slit die extrusion produces a flat film. Tubular film can be sliced to make flat
film. A typical process to produce blown or tubular film is shown in Figure 2.6.
Molten plastic from prior extruding enters the die where it is forced around the
mandrel (upward in Figure 2.6) and emerges from a ring shaped die. Air pressure

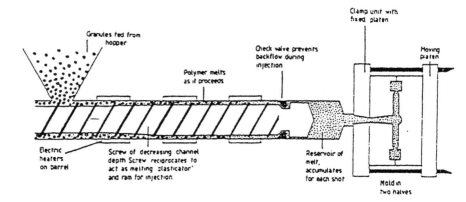

Figure 2.5   Simple injection mold machine [Morton-Jones, 1989]. (From Hegberg, B. A., et al., *Mixed Plastics Recycling Technology,* Noyes Data Corporation, Park Ridge, NJ, 1992. With permission.)

expands the formed tube to a predetermined diameter. The film thickness decreases during the expansion. The "frost line" shown in Figure 2.6 is the level at which solidification occurs.

In slit die or flat film extrusion, flat film is produced when molten plastic is extruded through a slit die and then cooled on chilled rollers or in a water bath

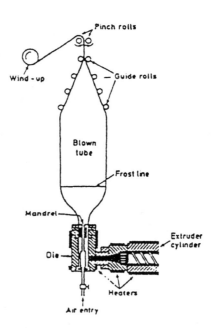

Figure 2.6   Blown film extrusion of tubular film [Briston, 1989]. (From Hegberg, B. A., et al., *Mixed Plastics Recycling Technology,* Noyes Data Corporation, Park Ridge, NJ, 1992. With permission.)

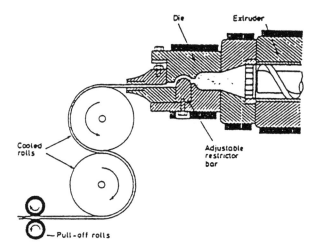

Figure 2.7 Slit die extrusion of flat film [Briston, 1989]. (From Hegberg, B. A., et al., *Mixed Plastics Recycling Technology,* Noyes Data Corporation, Park Ridge, NJ, 1992. With permission.)

as shown in Figure 2.7. It is important that cooling occurs close to the exit from the die in order to prevent reductions in film width.

It is important in the use of any extrusion process to maintain proper Melt Flow Index (MFI), a measure related to the viscosity of the melted plastic for a given resin. Table 2.16 lists density and MFI data[6] for four major resin types with

Table 2.16 Densities and Melt Flow Indexes of Production Polymers

Resin type	Density (g/cm³)	Melt flow index (g/10 min)
LLDPE film	0.918–0.927	0.1–2.5
LLDPE injection molding	0.915–0.928	5–400
LLDPE pipe and sheet	0.928–0.942	0.8–2.0
HDPE injection molding	0.958–0.968	5–400
HDPE blow molding	0.956–0.964	0.01–0.2
HDPE film	0.946–0.955	0.02–0.3
HDPE pipe and sheet	0.946–0.969	0.3–3.0
PP sheet extrusion	0.902–0.903	0.4–0.8
PP injection mold, general purpose	0.903	4–12
PP injection mold, thin complex parts	0.902	35
PS easy flow injection mold	1.04–1.08	16
PS medium flow injection mold	1.04–1.08	7.5
PS high heat injection mold	1.04–1.08	1.6

From Hegberg, B. A., et al., *Mixed Plastics Recycling Technology,* Noyes Data Corporation, Park Ridge, NJ, 1992. With permission.

three or four product production classifications for each resin type. The MFI determines how quickly the molten plastic will fill a die in terms of volumetric flow rate (gal/10 min). LLDPE (Linear Low Density Polyethylene) has very low viscosity with an MFI ranging between 5 and 400 depending on the viscosity and product/process application. HDPE used in blow molding lies at the other extreme with an MFI ranging from 0.01 to 0.2.

Resins used in blow molding retain a putty-like quality so that they will maintain thickness while being pressurized in a mold. Viscosity differences restrict the options available in using recycled plastics since, if different resins are comingled, they will not form a homogeneous melt, thus causing variations in MFI with resulting variations in formed product uniformity. There are no automated mechanical methods of separating different plastics based on resin types. If differences in density are large enough such as between PET and HDPE, water flotation will permit skimming HDPE off the surface while the PET will sink and be recovered from the bottom of the tank.

Plastic Lumber from Mixed Plastics

Plastic lumber is a very popular product derived from the secondary recycling of plastics. Plastic lumber is a flow molded linear product with a great number of uses and is also a product that can be made from mixed plastics even after the removal of HDPE and PET containers. The typical large cross-sections used in plastic lumber make it possible to use mixtures of resin types not normally considered compatible. For products requiring lighter colors such as yellow, orange, blue, or light gray, color separation is necessary. HDPE or LDPE that is clear or white must also be used to make light colored products. If a large portion of LDPE is used, the resulting material will be very elastic. A large portion of polypropylene will result in brittle plastic.

Blending of granulated recycled plastics made from specific resin types is important and must be controlled if product specifications are to be met. When separation by color is not employed, the resulting plastic lumber will tend to be dark brown, black, or dark gray. The economics of plastic lumber production and marketing are sensitive to the proximity of a recycled plastic processing facility, local plastic product manufacturers and affordable transportation. In some cases, it may be necessary to pay a manufacturer of products derived from recycled plastic to take the waste plastic. Overall, the market for products made from recycled plastic is expanding and is expected to grow steadily for a decade or more.

Manufacturers of Plastic Lumber

Table 2.17 lists nine companies that currently manufacture plastic lumber from mixed plastic waste.[6] Since the market for plastic lumber products is growing, more companies are entering the business. Three of the companies listed are in the northeastern U.S.; three are located in the Midwest, two in the south and one in the west. Companies taking only HDPE are not included in the table.

Table 2.17 Mixed Plastic Lumber Manufacturers

Company	Address	Plastic accepted	Comments
American Plastics Recycling Group	P.O. Box 68 Ionia, MI 48846 (616) 527-6677	65% Post-consumer 35% Industrial scrap	All rigid plastic containers; accepts films separated out and separated clear and colored HDPE; will accept mixed bales if HDPE bales go to them also, but not alone
Hammer's Plastic Recycling Corp.	RR. 3, Box 182 Iowa Falls, IA (515) 648-5073	60% Post-commercial 20% Industrial scrap 20% Post-consumer	All rigid plastic containers; reluctant to accept film unless separated from rest; community must take product fabricated back; mixed plastic may have to be separated for product
Innovative Plastic Products	P.O. Box 898 Greensboro, GA 30642 (404) 453-7552	Industrial scrap	Accepts only packaging and film scrap from industry; full start-up in early 1991
National Waste Technologies	67 Wall St. New York, NY 10005 (212) 323-8045	99% Post-consumer	
The Plastic Lumber Co.	209 S. Main Akron, OH 44308 (216) 762-8988	Post-consumer Industrial scrap Post-commercial	HDPE rigid plastics and LDPE films separated, washed and ground
Plastic Recyclers	58 Brook St. Bayshore, NY 11706 (516) 666-8700	50% Post-consumer 50% Industrial scrap	
Superwood Ontario	2430 Lucknow Mississauga ONT, CA Unit #1 5S1V3 (416) 672-3008	90+% Post-consumer 10% Industrial scrap	Any plastic accepted; containers and films should be washed out and rinsed; no separation is necessary; future suppliers may have to buy product back
Superwood Alabama	P.O. Box 2399 Selma, AL 36702-2399 (205) 874-3781	40% Post-consumer 60% Industrial scrap	Accepts separated clear and colored HDPE, and PP (must be separated); separated out films also accepted
Polymerix c/o Trimax	2076 Fifth Ave. Ronkenkoma, NY 11779 (516) 471-7777	85% Post-consumer 15% Post-commercial	Accepts mixed plastic and/or HDPE, depending on production level and influx from existing accounts

From Hegberg, B. A., et al., *Mixed Plastics Recycling Technology,* Noyes Data Corporation, Park Ridge, NJ, 1992. With permission.

Types of plastic accepted and preparation requirements vary with the manufacturer. The physical and economic requirements of preparation for recycling through a given plastic lumber manufacturer must be worked out carefully for each local recycling program. Transportation costs are also critical to the success of the program.

Plastic Lumber Manufacturing Processes

Some of the companies listed in Table 2.17 have their own patented processing machinery, and others have purchased machines developed by others such as the ET-1 machine produced by Advanced Recycling Technologies, Ltd. in Belgium.[27] All of the processes used by companies listed in Table 2.17 are basically similar and are suitable for producing thick molded parts as used in park benches or pallets or long profile extrusions used in plastic lumber. The ET-1 machine can process virtually all post-consumer plastics. Grinding to a 1/4 in. chip size is necessary prior to processing. Plastic film or thin sheets need to be compressed and reformed into small granules to provide sufficient friction in the extruder.[27] The chips and granules are blended together along with desired additives and colorants. In addition to post-consumer plastic waste, plastics from packaging, automotive parts and some electrical components can also be used in this process.

The ET-1 machine includes a screw type extruder, a molding chamber, a part removal mechanism and various controls. It operates within a melting temperature range of 360–400°F. Plastics such as PET and PC that melt at high temperatures and non-plastic contaminants such as aluminum and copper become filler material impacted within the other melted resins.

Another process, known as the Klobbie Process was developed in the Netherlands and is licensed through Superwood Holdings PLC. There are plants located in the U.S. and Canada using this process. Feedstock includes HDPE, LDPE, and PP for producing plastic lumber with PET and ABS accepted in limited quantities. The addition of PVC results in inferior product unless used sparingly. PVC processed independently can be used to produce plastic lumber of good quality. Sources of plastics used in the Superwood process include scrap from manufacturers of plastic film, bags, toys, trays, and other domestic products, as well as from producers of beverage containers. Low grade resin pellets from plastic processors are also used. Collected plastics are separated and graded for various secondary products and are blended in horizontal mixers. Ferrous metal is removed magnetically before the plastic is extruded. The Klobbie Process also uses a screw type extruder rotating in a barrel at a speed sufficient to cause melting of the resin mix from heat generated by friction. The melted plastic is forced through an orifice into a steel mold. Fillers are used to produce desired mechanical and visual characteristics. The process is then changed from extrusion to flow molding. Typically, 10 molds of similar length but differing cross-sectional area and shape are mounted horizontally on a carousel in a tank of water. The molds are rotated about a horizontal axis. At the top portion of their rotation they are filled with molten plastic from the extruder and then cooled as they are submerged back into the water. Parts are ejected from molds pneumatically. The water is recirculated and cooled by a chiller. The Klobbie Process is described in detail by Mulligan[28] and Curlee.[24]

The typical Superwood version will process approximately 400 lbs per hour and costs approximately $2 million to purchase. Operating labor includes a primary operator with one or two assistants for the extruder, plus two or three sorters of infeed material and others as needed to fabricate end products.[28] Mackzo[27] lists a number of post-consumer and post-industrial plastics together with recommendations for their use in the Superwood machine as follows:

- *LDPE or LLDPE* A good material for use in the process. However, LDPE is relatively soft and products containing too much of it may be insufficiently rigid for some applications, particularly in thin sections. It should be mixed with stiffer materials such as HDPE or PP.

- *HDPE* A good material for use in the process. HDPE is stiff and its mixtures with LDPE give a range of stiffness that cover most product requirements. Much of the HDPE on the market is copolymer material, but this is of no consequence to the recycler because for recycling purposes its performance is very similar to that of a homopolymer.

- *PP* A good material for use in the process. It is relatively stiff and its mixtures with LDPE cover most of the range of stiffness requirements. However, the use of more than 30% by weight homopolymer PP is not advised because it is brittle at low temperatures and difficult to nail.

- *PVC* When finely ground and well homogenized PVC can be recycled on the ET-1. It can be mixed with other thermoplastics up to 50% by weight. Post-consumer plastic typically contains 5% PVC or less.

- *PS* Up to 40% by weight of this material can be mixed in. Impact grades add toughness to the mix. Non-impact PS (crystal) tends to cause surface finish problems. Expanded PS (EPS) should be avoided as a foam because of its low bulk density. Testing shows considerable strength improvements at 10 to 40% levels of densified EPS.

- *ABS* A good material for use in the process. The ABS family of resins combines rubbery and plastic properties and is extremely tough. ABS plastics are not broadly available.

- *Nylon* A wide variety is currently on the market. The most common, nylon 6 and 6/6, can be an additive at up to 10% by weight because they impart stiffness to an otherwise soft compound. Textile nylon scrap is usually nylon 6 or 6/6. Nylon 6 castings are suitable. Nylons 11 and 12 are even more suitable, but generally not available.

- *PET* Although its 500°F melt temperature is above the normal range of the ET-1, up to 15% can be mixed in if finely ground and carefully blended. PET beverage bottles, with HDPE base cup, labels, and aluminum caps have been run at 100%, but the product is brittle due to crystallization caused by slow cooling of thick sections and degradation of the polymer caused by moisture content.

Hammer's Plastic Recycling process is patented in the U.S. and was developed in Iowa where the owner company is located. This process is designed to accept post-consumer plastic as well as post-industrial and commercial.[29] A typical blended mixture for this process is 65% LDPE, 20% HDPE, 5% PP, 5% PET, and 5% miscellaneous. Closed molds are filled at pressures ranging from 100 to 600 psi. Hammer's process can operate at production rates of 800–1000 lbs per hour.

Properties of Plastic Lumber

Plastic lumber has a number of properties that make it superior to traditional lumber such as its resistance to damage from weather, insects, and vandalism. But, it also costs more and its advantage must be sold in the basis of cost effectiveness. The Center for Plastics Recycling Research at Rutgers University, NJ has performed research on the effect of adding various percentages of PS to plastics tailings (left after the removal of clear HDPE and PET bottles) in the compressive strength of the resulting plastic.[6] Compressive and yield strength increased approximately 15 and 20%, respectively, for each 10% addition of PS up to a 50% addition. Above 35% PS, the modulus of elasticity decreased slightly and reached a limit of about 220,000 psi. Table 2.18 shows the overall outcome of this research by Nosker, Renfree, and Morrow.[30]

Other tests reported by Mack[31] compared the static bending strength of extruded plastic lumber reinforced with glass fiber and a foamed plastic core to that of wet and dry Southern Yellow Pine, using standard sized "2 × 4" pieces of each material. The results are shown in Figure 2.8. Glass fiber reinforced, foam cored plastic lumber made from consumer scrap sustained nearly twice the load for a given deflection as compared with unreinforced, unfoamed plastic lumber. Wet Southern Yellow Pine had superior strength compared with plastic lumber

Table 2.18   Change in Mechanical Properties of Plastic Tailings with Addition of Polystyrene [Nosker et al., 1990]

Composition (%)		Compressive modulus (psi)	Yield stress (psi)	Compressive strength (psi)
0 PS	100 tailings	90,000	2,700	3,170
10 PS	90 tailings	144,370	3,100	3,220
20 PS	80 tailings	163,390	3,860	3,860
30 PS	70 tailings	197,600	4,350	4,350
35 PS	65 tailings	239,000	4,950	4,950
40 PS	60 tailings	222,300	4,750	4,750
50 PS	50 tailings	220,000	5,320	5,320
Standard virgin resins:[a]				
HDPE		90,000–150,000	—	2,500
LDPE		20,000–27,000	—	900–2,500
PP		100,000–170,000	—	4,000
PS		450,000	—	6,000–7,300

[a] From *Perry's Chemical Engineers Handbook,* Sixth Ed., McGraw-Hill, New York, 1994. Modulus in tension shown.

From Hegberg, B. A., et al., *Mixed Plastics Recycling Technology,* Noyes Data Corporation, Park Ridge, NJ, 1992. With permission.

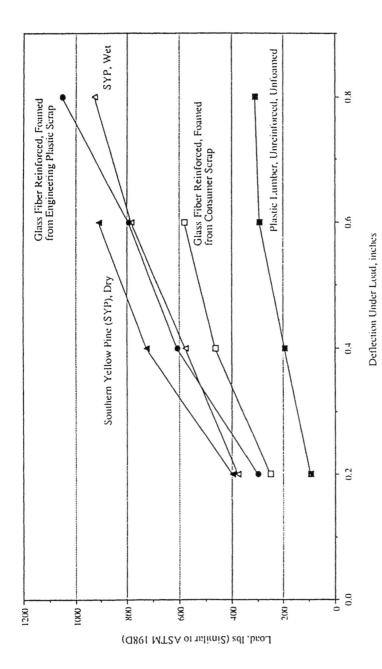

Figure 2.8 Static bending test results of 2 × 4 lumber pieces [Mack, 1990]. (From Hegberg, B. A., et al., *Mixed Plastics Recycling Technology*, Noyes Data Corporation, Park Ridge, NJ, 1992. With permission.)

derived from consumer scrap. Glass fiber reinforced, foam cored plastic derived from industrial waste compared favorably with wet Southern Yellow Pine but was not as resistant to deflection as dry Southern Yellow Pine.

Plastics have also been considered as reinforcing material for concrete. Researchers at the Oak Ridge National Laboratory in Tennessee have developed a new type of concrete called "superconcrete" that contains recycled plastic and is more resistant to erosion than ordinary concrete. The purpose of its development was to serve as containment for toxic waste. It could also serve as a lighter weight material for use on bridges or for paving highways. It would not develop potholes as readily as ordinary concrete. Plastics used in "superconcrete" include polystyrene (as used in coffee cups), a material that has caused environmental problems. Several companies have expressed an interest in manufacturing and marketing "superconcrete."

The addition of wood fiber to recycled plastic resin in the manufacture of plastic lumber has been found to improve its mechanical properties as well as its dimensional stability. The use of wood fibers as reinforcing material for plastic lumber is an important development since wood fiber is abundant, lightweight, and economical. Research has been performed to study the effects on mechanical properties of plastic lumber resulting from the addition of treated and untreated aspen fibers into plastic chips made from HDPE milk jugs.[32] Tests showed a reduction in tensile strength and Izod impact strength below values obtained using HDPE alone when aspen fiber was added. Tensile and flexural modulus values showed the reverse effect and were significantly higher than corresponding values for HDPE alone. Uniform dispersion of the fiber in the resin is important in achieving improvements in tensile and flexural modulus.

Additional work has been done in studying the effects of using recycled wood fiber–polystryene composite in the same composite material on mechanical properties of the resulting material under extreme temperatures (–20°C to 105°C). Mechanical properties and dimensional stability did not change significantly between original and recycled composite material.[33]

Summary

The market for plastic lumber products is growing steadily. What is lacking is an integrated infrastructure for collecting, processing, transporting, and manufacturing plastic lumber on a regional basis at prices that are competitive with those of virgin materials. Since it is generally uneconomical to transport low density, unground plastic containers or materials more than 100–200 miles, concentrating a recycling, re-manufacturing effort in a large metropolitan area is more likely to assure its financial viability.

EMERGING TECHNOLOGY FOR SEPARATION AND PROCESSING

Pre-sorting by the consumer and manual sorting by employees in a material recycling facility (MRF) still remain as the most widely used methods for separating different types of plastics.

The low productivity of this type of manual labor and the generally low prices received for recycled plastics have stimulated interest in developing automated sorting techniques. Automated sorting technology has relied upon differentiation in basic physical properties among the major types of resins according to optical recognition (opacity), X-ray diffraction, solubility in solvents, and specific gravity. None of these methods is effective in differentiating among all six basic resin types accounting for most post-consumer plastic. Research in automated classification can be defined as being on a macro, micro, or molecular scale.[6] Macro sorting means separating plastics based on scanning an entire product such as occurs when optical sensing is used to separate plastic jugs by color; manual sorting is a form of macro sorting. Micro sorting is done on chips or flakes after the products are ground up. This is frequently done in separating HDPE and PET as is necessary for plastic bottles having an HDPE base cup with a PET upper chamber. In a flotation tank or hydroclone, PET will sink in water while HDPE floats. Molecular separation requires that the plastics be dissolved in a solvent and then separated based on temperature ranges.

Separating PVC Bottles

It is particularly important to remove PVC bottles from other plastic products because of its degrading effects on other common plastics such as PET. Another unwanted effect occurs when the chlorine released from the melted PVC combines with hydrogen from the plastic polymers to form hydrochloric acid, which corrodes extruder dies and other components.[6]

Research on separating plastics by the B.F. Goodrich Company and the Rutgers University Center for Plastics Recycling Research has been done using X-ray fluorescence to detect the scattering effect caused by chorine atoms in PVC. Mechanical separation is then carried out, based on the X-ray detection of PVC.[34] Other plastics containing chlorine such as polyvinylidene chloride (PVDC) used in laminating and in making plastic films can also be detected using X-ray fluorescence. While effective in detecting small amounts of chlorine, the X-ray that was used does not penetrate paper labels and is extremely sensitive to distance from the X-ray source to the target container, weakening rapidly as the sample is moved away from the X-ray source and detector. This effect could cause problems when bottles or jugs with highly variable shapes or orientations are being processed.[35] The strength of this signal indicating chlorine decreases to half of its value with the addition of each 3 mm of air space between the sample being studied and the detecting device. This research study also showed that the signal intensity for chlorine detection also varied along the length of the bottle. A 2 cm interval was found to be optimum for maximizing the signal strength.[36] The X-ray process used in this study was capable of generating a signal every 0.005 sec. For a sampling interval of 2 cm, this suggests a linear velocity of 400 cm/sec, providing the capability of scanning 10 bottles per second with a 10 cm spacing between the bottles.[6]

Research by the same authors involving an actual plastics recycling process in Akron, Ohio showed manual separation of PVC bottles to be only about 80%

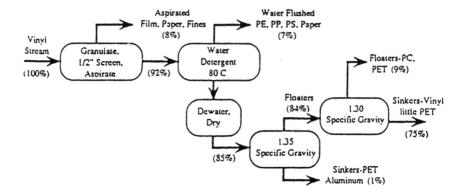

Figure 2.9 Vinyl stream purification process [Summers et al., 1990b]. (From Hegberg, B. A., et al., *Mixed Plastics Recycling Technology,* Noyes Data Corporation, Park Ridge, NJ, 1992. With permission.)

efficient in the identification and removal of PVC bottles from the mixed plastic waste stream.[36] When automated detection and separation using X-rays was employed, additional cleaning was found to be necessary. A method was proposed for cleaning and separating using solutions of calcium nitrate with specific gravity values of 1.3 and 1.35. The overall process is shown in Figure 2.9.

Separating HDPE Base Cups from PET Beverage Bottles

A process has been developed and demonstrated at the Center for Plastics Recycling Research (CPRR) at Rutgers University.[37,38] This process known as the Rutgers Beverage Bottle Reclamation Process (BBRP) has been made available for licensing and royalty fees from Rutgers University.[6] Other commercial processes exist for recycling PET beverage bottles, but their technologies are proprietary company information and not available for licensing. The Rutgers BBRP does not separate PET bottles by color or from other plastics. It is designed to separate cleaned PET as flakes from plastic or aluminum caps, paper labels, HDPE base cup and adhesives used to attach the label or the base cup to the main container wall. Materials used in this type of bottle vary. A representative average PET bottle composition is shown in Table 2.19. The HDPE base cups are also ground into flakes that include the PP caps and fragments of aluminum caps. The overall process is diagrammed in Figure 2.10[19] and described as follows:[37,38]

- Collection of waste plastic.
- Sorting of waste plastic into uniform types.
- Granulation or cutting of the sorted plastics into chips of about 3/8 in. mesh size. This may be accompanied by air classification to remove loose dirt, paper shreds, and other "fines."
- Mixing of the plastic chips with a heated aqueous detergent bath, and agitating the resulting slurry for sufficient time to achieve the desired degree of cleanliness, including label removal, EVA removal and disintegration of any paper present.

Table 2.19   Average Composition of Mixed Soda Bottles Received at Rutgers, NJ During 1988–1989 [Dittman, 1990]

Compound	Average compostition (weight %)
PET (including green and clear)	73.6
HDPE[a]	20.0
Paper (label)[b]	2.9
EVA[c]	1.3
Aluminum	1.0
PP (labels)[d]	0.8
PP (cap liner)	0.4

[a] Actual HDPE quantity is 23% for bottles with base cups and 0% without base cups.
[b] Value goes to 0% for bottles with plastic labels.
[c] Actual ethylene–vinyl acetate (EVA) quantity, which is a base cup adhesive, is 1% for bottles without base cups and 3% for bottles with base cups. EVA is a copolymer adhesive in the polyethylene.
[d] Value goes to 0.2% for bottles with paper labels. For those bottles, PP is used for cap liners only.

From Hegberg, B. A., et al., *Mixed Plastics Recycling Technology,* Noyes Data Corporation, Park Ridge, NJ, 1992. With permission.

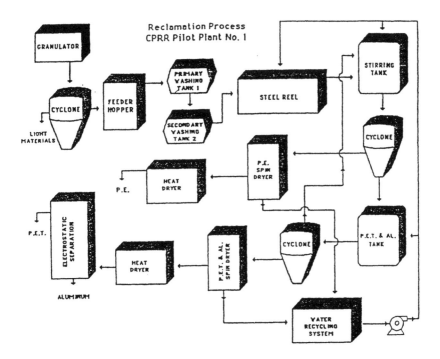

Figure 2.10   Schematic of Rutgers Beverage Bottle Reclamation Process [Morrow and Merriam, 1989]. (From Hegberg, B. A., et al., *Mixed Plastics Recycling Technology,* Noyes Data Corporation, Park Ridge, NJ, 1992. With permission.)

- Separation of the wash bath from the plastic chips via a dewatering screening device. Paper fibers and dirt are removed along with the bath; except for the portion which remain clinging to the wet chips. The wash bath is then filtered and reused. Make-up detergent solution is added as required.
- The wet and washed chips are re-slurried in rinse water in an agitated tank, and pumped through a set of hydrocyclones to separate the "light" components (polyethylene base cups and polypropylene labels) from the "heavy" components (PET and aluminum bottle tops). The specific gravities are as follows:

   a. HDPE 0.96
   b. PP   0.90
   c. PET   1.29 to 1.40
   d. Al   2.60

Thus, items a and b will float in water, while c and d will sink.
- The light and heavy components are individually dewatered to 3–7% H_2O using high-speed rotating fine screens and dried. Thermal separation is used to remove residual water from the two fractions after dewatering. One dryer is used for each stream, with the PE/PP stream dried to 1% H_2O and the PET/Al stream dried to 0.179% H_2O or less. The low moisture content of the latter is necessary for electrostatic removal of the aluminum.
- The light component stream is then ready for sale, or an optional step can be used to remove PP from PE. PE/PP separation can be done by increasing the air velocity in the light stream hot-air-drier. PP chips will be carried out by the air. It has been found, however, that most purchasers are satisfied to buy HDPE chips containing 7–8% PP by weight. They can be used as a mixture in injection molding, since their melting point temperatures are not far apart:

HDPE 130–137°C
PP   168–175°C

- The heavy component stream is sent to an electrostatic Carpco separator for removing aluminum. The PET and Al mixture is fed continuously onto a rotating steel roll maintained at high voltage. The electrostatic aluminum removal process takes aluminum from 1.2% (12,000 ppm) to 50–100 ppm residual aluminum in the cleaned PET flake. The aluminum by-product contains approximately 50% Al and 50% PET by weight. An additional smaller separator is necessary to attain a higher purity aluminum material.

Analysis of the output of the Rutgers BBRP showed compositions as listed in Table 2.20. Wastewater from the process has not caused environmental problems. Waste by-products are reported to consist primarily of wet sludge composed of dirty water and paper fibers. Rutgers University has installed a 5 million pound per year capacity pilot plant to demonstrate the overall process.[37]

A cost estimate for a 20 million pound (10,000 ton) per year processing plant for recycled plastics was calculated by the Rutgers CPRR along with an estimated rate of returns on investment assuming market prices for clean flaked PET at levels of $0.31, 0.36, and 0.41 per lb. At these prices, the estimated return on

Table 2.20 Product Analysis of Rutgers Beverage Bottle
Reclamation Process [Dittman, 1990]

Component	Mean PET product	Mean HDPE product
PET, including EVA (wt%)	99.756%	0.02 to 0.63%
PET, EVA free basis[a] (wt%)	99.995%	—
HDPE	10.3 ppm	92.4–99.4%
PP	1.2 ppm	0.0–7.5 ppm
Aluminum	25–100 ppm	0.0–2.0 ppm
Average moisture content	≤0.25%	1%

[a] Ethylene–vinyl acetate (EVA) can be removed to as high a
degree as desired.

From Hegberg, B. A., et al., *Mixed Plastics Recycling Technology,* Noyes Data Corporation, Park Ridge, NJ, 1992. With permission.

investment was estimated at 34, 50, and 66% per year, respectively for these three
market prices. A similar analysis for a 10 million pound (5000 ton) per year
process for the same prices resulted in annual rates of return of 4, 16, and 29%,
respectively.[39]

Separation by Selective Dissolution

This method of separating plastics on a molecular scale is based on the
selective removal of different resins in a common solvent.[6] Research in this
method began in the mid-1970s and has continued as an experimental approach
used in laboratories. Two approaches are used. In the first, several resin types are
dissolved in a common solvent. In the second, a solvent that dissolves one resin
is used to remove that particular resin, leaving the others as undissolved residue.
Both methods have potential value since not only is the plastic waste stream
heterogeneous, but it also contains other solid waste materials. Selective dissolution can be very thorough in eliminating or dispersing contaminants that can
cause manufacturing problems or result in poor physical properties for the recon-
stituted plastic.

A two-stage process is required to obtain solvent-free polymer:[40]

- Selective dissolution — A solvent and a sequence of solvation temperatures are
used in a sequential batch mode to selectively extract a single polymer group
from the comingled stream. The polymer obtained from the single extraction
is isolated using flash devolatilization. The recovered polymer is then pelletized
and the condensed solvent is returned to the dissolution reservoir to remove the
next group of polymers at a higher temperature. Typically conditions consisted
of placing 25 kg of plastic waste (virgin polymers used) into the column with
screens at each end. A pump circulated 20 liters of solvent through the heat
exchanger and dissolution column.
- Flash devolatilization — Once separated by selective dissolution, flash devolatilization of the solvent form the mixture is used to produce solvent-free

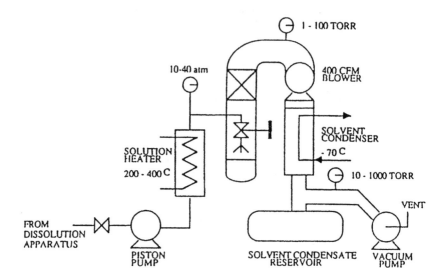

Figure 2.11   Selective dissolution single solvent flash devolatilization mini-plant [Lynch and Nauman, 1989]. (From Hegberg, B. A., et al., *Mixed Plastics Recycling Technology,* Noyes Data Corporation, Park Ridge, NJ, 1992. With permission.)

polymers. This process is an outgrowth of the polymer process of compositional quenching, where two incompatible polymers are dissolved in a common solvent, and then the solvent devolatilized such that phase separation between the two polymers occurs and a microdispersion of one polymer within the other occurs. The microdispersion renders the minor constituents innocuous. Figure 2.11 shows an experimental flash devolatilization apparatus. The polymer concentrations in the solvent are typically 5–10% by weight. The pressure in the heat exchanger is sufficient to prevent boiling with pressure maintained by the flash valve at 10–40 atm, and typical temperature upstream of the flash valve and the flash chamber pressure (typically 5–100 torr) are the two variables which govern the devolatilization step. They determine the polymer concentration after flashing (typically 60–95%) and the after flash temperature, 0–100°C.

The overall flowchart for this process is shown in Figure 2.12. Tests of this system were conducted using equal volumes of six common thermoplastics: LDPE, HDPE, PET, PP, PS, and PVC. Tetrahydrofuran was used as the solvent. Xylene has also been used as a solvent for these plastics. Results of the tests indicated a recovery efficiency of over 99% for all six thermoplastics at temperatures ranging from 25°C for PVC and PS to 190°C for PET. PVC and PS were separated first at 25°C second at 70°C, PP and HDPE third at 160°C and PET fourth at 190°C.[40] The cost of processing plastics by this method was estimated at $0.15 per pound for a 50 million pound (25,000 ton) per year process.

A multiple solvent process ran also be used. In this method a different solvent is used for each polymer, yielding purer plastics than can usually be obtained using a single solvent. The multiple solvent process has been used to purify PET flakes after cleaning and to separate them from HDPE, PP, paper, and aluminum.

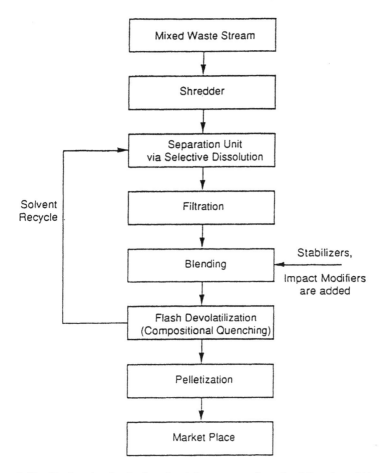

Figure 2.12   Single solvent selective dissolution process flow sheet [Lynch and Nauman, 1989]. (From Hegberg, B. A., et al., *Mixed Plastics Recycling Technology*, Noyes Data Corporation, Park Ridge, NJ, 1992. With permission.)

This purification entails an additional cost beyond that required for ordinary processing using a single solvent. Detailed information on this type of process is available.[44]

Separation Using Soluble Acrylic Polymers

The Belland AG company located in Switzerland specializes in the development and use of acrylic polymers. These are suitable for use as solvents for separating acrylic plastics. Belland AG has developed processes suitable for separating acrylic polymers which contain carboxyl additives, which cause the polymers to be soluble in an alkaline solution.[6] Precipitation of the polymers can be achieved by reducing the pH of the solution through the addition of an acid. Detailed information on this technology and its application is available.[45]

Dissolution of Shredded Automobile Plastic

Although metals constitute the primary recovered materials from the shredding of automobiles, plastics also are a significant recoverable constituent. It has been estimated that a 2500 lb automobile contains between 250 and 275 lb of plastic and that up to 1.2 million tons of waste plastic are generated each year from scrapped automobiles.[46] In this case, contaminants include motor oil, power steering and brake fluids, grease, and possibly heavy metals. In terms of composition, thermoplastics account for 65% of the total plastic waste. These include PUR, PP, PVC, and ABS. The Argonne National Laboratories have conducted research to improve the recovery of plastics from shredded automobiles. Automotive companies have also performed studies involving the pyrolysis of thermoset plastics (heating them in an oxygen free atmosphere). When volatile constituents are evaporated in this process, the resulting products are combustible gas or fuel oil. These can be used to provide heat for the process or as a filler material in manufacturing new molded plastic.

Degradable Plastics

There are presently six methods of producing or quickening degradation of plastics.[5] The most viable in terms of potential commercial development are: photodegradation, biodegradation and biodeterioration. The others are: auto-oxidation, hydrolysis and solubilization. These six processes occur by the following mechanisms of deterioration:

Photodegradation — reactions within the polymer caused by sunlight.
Biodegradation — decomposition caused by bacteria, algae, fungi, or yeast.
Biodeterioration — decomposition caused by insects such as beetles or by slugs.
Auto-oxidation — degradation caused by combining of plastic molecules with oxygen.
Hydrolysis — breakdown of polymer chains by water causing a decrease in molecular weight and physical strength characteristics.
Solubilization — dissolution of polymers resulting from the presence of water-soluble links in the polymer. The polymeric structural form remains.

Precise definitions and measures of performance for these processes are still lacking. This has been cited as a reason for the slow rate of development of these plastics.[47] The American Society for Testing and Materials (ASTM) is developing standards for testing and measuring the degradability of plastics in terms of changes in their physical properties over time (e.g., losing tensile strength or becoming brittle). A list of companies that have developed degradation processes is given in Table 2.21.[48]

Photodegradation and biodegradation depend on the environmental conditions to which the given plastic is exposed as well as upon the composition of the plastic. The former depends on reactions caused by ultraviolet radiation within

Table 2.21   Summary of Degradation Rates for Available Technologies

Developer	Product sold	Manufacturers reported time to degradation[a]	Characteristics of product degraded
Ecoplastics Willowdale, ONT	Photodegradable ketone caroonyl copolymers	Not available	—
DuPont Co. Wilmington, DE	Photodegradable ethylene/carbon monoxide copolymer	4–5 days. Calif. in summer, 60 days in Alaska in fall	LDPE polymer with 1% CO copolymer
Union Carbide Danbury, CT	Photodegradable ethylene/carbon monoxide copolymer	60 days in New Jersey in winter	LDPE polymer with 2.7% CO copolymer
Ampacet Mt. Vernon, NY	Photodegradable additive system	8–28 wk at varied U.S. locations and seasons	LDPE film with "Polygrade" masterbatch
Ideamasters Miami, FL/Israel	Photodegradable additive system	3 wks — Israel test; 48 wks — European test	—
ICI Americas Wilmington, DE	Biodegradable aliphatic polyester copolymer	Case 1 — in a matter of days in sewage treatment plant; Case 2 — 1 yr (est.)	Case 1 — Thin film; Case 2 — Bottle
St. Lawrence Starch Mississauga, ONT	Biodegradable starch additive	3–6 yr in sanitary landfill (est.)	Resin with 6% starch

Note: Not all of these technologies are currently available commercially.

[a] The meaning of "degradation" in the reports cited is varied.

Compiled by Eastern Research Group from Society of the Plastics Industry, 1987, as published by Curlee, T. R. and Das, S., *Plastic Wastes: Management, Control, Recycling and Disposal*, Noyes Data Corporation, Park Ridge, NJ, 1991. With permission.

photosensitive material. Specifically, this involves the breaking up of long polymeric molecular chains. Direct sunlight is needed to produce photodegradation, requiring it to occur outdoors. Thus, plastic products used indoors can be disposed of by this method when their functional life cycle is complete.

Photodegradation is generally made to occur through the addition of carbon monoxide, resulting in the formation of carbonyl groups within the polymer chains. Upon absorbing a sufficient amount of ultraviolet light linkages breakdown in the polymer chains. A number of commercially available additives have been developed for enhancing photodegradability. One developed at the Princeton Polymer Laboratory,[48] uses a photsensitizing agent combined with a metallic compound. Other methods employ antioxidants at low concentration levels to accelerate photodegradation.

Biodegradation in plastics is generally brought about either by modification of the polymer or by the addition of special chemical agents. When the latter method is employed, the polymer remains but may be altered from its original form.[5] Some polyester and polyurethane resins are biodegradable without the addition of special chemical agents. These are used in agricultural applications (e.g., seedling containers) and do not become part of the plastic waste stream. Starch has been a popular chemical additive, being readily available, inexpensive, and very biodegradable through bacterial action in soil. Degradation of the remaining polymer is not accelerated through the use of starch as an additive. Starch may be added in a 5 to 10% ratio relative to the polymer resin. The U.S. Department of Agriculture has conducted experiments using high starch concentrations. In this case, the starch was gelatinized before being mixed into the polymer. The result was biodegradable starch and a porous plastic of lower molecular weight making it more biodegradable also.[50]

When auto-oxidation agents are added to polymers, these react with metals in the soil when the plastic product is buried. Hydrolysis requires that water react with the polymer chains, causing them to break down and have a lower molecular weight. It is necessary for water-reactive chemical groups be present in the polymer for this effect to occur. Ester groups are effective in this role.

The U.S. Food and Drug Administration (FDA) should be contacted regarding the use of any degradable plastic for food packaging or other applications involving contact with food products. Concerns of the FDA include: the conditions and time interval in which biodegradation occurs; environmental impacts; effects of plastic fragments on ecosystems; and impact on recycling programs.[47] Other environmental concerns include possible relationships between toxicity and biodegradability and leaking of chemical additives into the soil.

Promoters of recycling programs have stated that the use of degradable plastics could complicate recycling by degrading the overall quality of mixed recycled plastics if their concentration is sufficiently high. This issue has yet to be resolved. In the case of yard waste, biodegradable plastic bags are desirable since their decomposition would be compatible with the composting process.

SUMMARY

Emerging technology for separating and processing recycled plastics has followed several technological paths and is continuing to grow. Financial incentives to develop improved processes have not been as strong as desirable if widespread, cost effective recycling is to occur in the near future. Collection, transportation, separation, and processing are all major issues affecting the economic viability of recycling. Economies of scale are important and currently favor only the most physically efficient processes conducted on an extremely large scale and feasible in only a few large metropolitan areas. Degradable and nondegradable plastics present different and, in some cases, conflicting issues in recycling. A total systems approach simultaneously recognizing physical, economic, and environmental objectives has yet to be developed. Cooperative efforts involving industry, local and federal governmental agencies, and universities should be promoted in dealing with the many technical and economic issues involved in plastics recycling.

REFERENCES

1. Adams, S. K., Even, J. C., Jr., Gupta, S., Wang, C. H., and Burroughs, B., *A Communications Network for Recyclable Materials,* Year 1 Report, 1991. Sponsored by U.S. Environmental Protection Agency, Region VII, Kansas City, KS.
2. Adams, S. K., Even, J. C., Jr., Loete, J. L., Kim, K. J., and Oberbroeckling, G. L., *Seasonal Flow Stream Analysis of the City of Ames Resource Recovery Plant,* 1991. Sponsored by the City of Ames, IA.
3. Adams, S. K., Even, J. C., Jr., Wagner, C., and Wellman, M., *Regionalized Collection, Processing, Material Recycling, and Energy Derivation from Solid Waste,* ISU-ERI-Ames 90002, Project 3042, 1989, Sponsored by the Iowa Department of Natural Resources through the Iowa State University Water Resources Institute, Grant No. C7:87-G280-01 Project 3042.
4. Coghlan, A., There's Oil in Them There Plastics, *New Scientist,* December 11, 1993, pp. 22–23.
5. Curlee, T. R. and Das, S. *Plastic Wastes: Management, Control, Recycling and Disposal,* Noyes Data Corp., Park Ridge, NJ, 1991, (a) Part 1: Wastes in the Municipal Solid Waste Stream and (b) Part 2: Recycling in the Industrial Sector.
6. Hegberg, B. A., Brenniman, G. R., and Hallenbeck, W. H., *Mixed Plastics Recycling Technology,* Noyes Data Corp., Park Ridge, NJ, 1992.
7. [Anon] New Ways to Recycle Plastic, *Mechanical Engineering,* June, 1994, p. 30.
8. U.S. Environmental Protection Agency, *The Solid Waste Dilemma: An Agenda for Action,* Publication No. EPA/530-SW-89-019. U.S. EPA Office of Solid Waste, Washington, D.C., 1989.
9. U.S. Environmental Protection Agency, *Methods to Manage and Control Plastic Wastes — Report to Congress,* Publication No. EPA/530-SW-89-051, U.S. EPA Office of Solid Waste, Washington, D.C., 1990a.
10. U.S. Environmental Protection Agency, *Characterization of Municipal Solid Waste in the U.S.: 1990 Update,* Publication No. EPA/530-SW-90-042A. U.S. EPA Office of Solid Waste, Washington, D.C., 1990b.
11. Shut, J., A Barrage of News from the Recycling Front, *Plastics Technology,* 36 (7), pp. 109–119, 1990.
12. Plastics News, PET Recycling Hits All Time High in 1989, *Plastics News,* 2 (36), November 2, 1990.
13. Modern Plastics, U.S. Resin Sales, *Modern Plastics,* 67 (1), pp. 99–109, 1990.
14. Harrison, S., Closing the Loop In-House, *Waste Age,* 26 (7), pp. 79–83, 1995.
15. Butcher, S. W., Maximizing Recycling Rates: Participation Versus Efficiency, *Waste Age,* 26 (7), July, 1995, pp. 85–91.
16. Franklin Associates, *Characterization of Plastic Products in Municipal Solid Waste — Final Report,* Prepared for Council for Solid Waste Solutions, Prairie Village, KS, 1990.
17. Peritz, L., Vice President, wTe Corp., Plastic Processing Technology Case Study, *Presentation at the 1990 National Recycling Congress,* San Diego, CA, 1990.
18. City of Chicago (Draft) *Solid Waste Management Plan Waste Characterization Report for Chicago, Illinois,* HDR Engineering, Inc., Omaha, NE, 1990.
19. Morrow, D. and Merriam, G., Recycling — Collection Systems for Plastics in Municipal Solid Wastes — A New Raw Material, *Proceedings of the Society of Plastics Engineers RETEC Conference — New Developments in Plastics Recycling,* Charlotte, NC, 1989.

20. Rankin, S., Ed., Frankel, H., Hanesian, D., Merriam, C., Nosker, T. and Roche, E., *Plastics Collection and Sorting: Including Plastics in a Multi-Material Recycling Program for Non-Rural Single Family Homes*, Center for Plastics Recycling Research, Rutgers, The State University of New Jersey, Piscataway, NJ, 1988.

21. CSWS (Council for Solid Waste Solutions), Plastics Recycling Pilot Program Final Report: Findings and Conclusions — Prepared for Hennepin County, MN, Washington, D.C., 1990.

22. Feaincombe, J., *Guide for Recyclers of Plastic Packaging in Illinois*, Prepared by Bottom Line Consulting, Inc. for the Illinois Department of Energy and Natural Resources, Springfield, IL, 1990.

23. Center for Plastics Recycling Research, *Plastics Collection and Sorting: Including Plastics in a Multi-Material Recycling Program for Non-rural Single Family Homes*, Rutgers University, Piscataway, NJ, 1988.

24. Curlee, T. R., *The Economic Feasibility of Recycling: A Case Study of Plastic Wastes*, Praeger Publishing Co., New York, NY, 1986.

25. McMurrier, M., Assessing a Polymer's Recyclability, *Plastic Recycling Machinery and Equipment Report*, (Supplement to Plastics Machinery and Equipment and Plastics Compounding), September 1990, cited in Reference 6.

26. Miller, E., *Plastic Products Design Handbook — Part B: Process and Design for Processes*, Marcel Dekker, New York, NY, 1983.

27. Mackzo, J., An Alternative to Landfills for Mixed Plastic Waste, *Plastics Engineering,* 46 (4), pp. 51–53, 1990.

28. Mulligan, T., Commercially Viable Products from Plastic Waste — The Superwood Process, *Proceedings of the Society of Plastics Engineers RETEC Conference — New Developments in Plastics Recycling*, Charlotte, NC, 1989.

29. Hammer, F., Production and Marketing of Products from Mixed Plastic Waste, *Proceedings of the Society of Plastics Engineers RETEC Conference — New Developments in Plastics Recycling*, Charlotte, NC, 1989.

30. Nosker, T., Renfree, R., and Morrow, D., Recycle Polystyrene Add Value to Comingled Products, *Plastics Engineering,* 46 (2), pp. 33–36, 1990.

31. Mack, W., Turning Mixed Plastic Waste into Specification Products through Advanced Technology, *Proceedings of the Society of Plastics Engineers Regional Technical Conference — Recycling Technology of the '90's*, Chicago, IL, 1990.

32. Yam, K., Gogoi, B., Lai, C., and Selke, B., Composites from Compounding Wood Fibers with Recycled High Density Polyethylene, *Polymer Engineering and Science,* 30 (11), pp. 693–699, 1990.

33. Maldas, D. and Kokta, B., Effect of Recycling on the Mechanical Properties of Wood Fiber-Polystyrene Composites, Part I, *Polymer Composites,* 11 (2), pp. 77–83, 1990.

34. Monks, R. New Reclaim Methods Target PVC, *Plastics Technology,* 36 (5) pp. 31–32, 1990.

35. Summers, J., Mikofalvy, B., and Little, S. Use of X-ray Fluorescence for Sorting Vinyl from Other Packaging Materials in Municipal Solid Waste *Journal of Vinyl Technology,* 12 (3), pp. 161–164, 1990.

36. Summers, J., Mikofalvy, B., Wooten, B., and Sell, W. Recycling Vinyl Packaging Materials from the City of Akron Municipal Wastes, *Journal of Vinyl Technology,* 12 (3), pp. 154–160, 1990.

37. Dittman, F. New Developments in the Processing of Recycled Plastics, *Proceedings of the Society of Plastics Engineers Regional Technical Conference — Recycling Technology of the '90's*, Chicago, IL, 1990.

38. Rankon, S. *Plastics Recycling Processes,* Rutgers Center for Plastics Recycling Research, Piscataway, NJ, 1990.
39. Phillips, E. and Alex, A. The Economics of Reclamation of Generic Plastics from Post-Consumer Wastes, Presented at the American Chemical Society MARM (Middle Atlantic Regional Meeting) Conference, Florham Park, NJ, 1990.
40. Lynch, J. and Nauman, E. Separation of Comingled Plastics by Selective Dissolution, *Proceedings of the Society of Plastics Engineers RETEC Conference — New Developments in Plastics Recycling,* Charlotte, NC, 1989.
41. Miller, E. *Plastic Products Design Handbook — Part B: Process Design for Processes.* Marcel Dekker, New York, NY, 1983.
42. Briston, J. *Plastic Films,* 3rd Ed. Longman Scientific and Technical, Essex, U.K., 1989.
43. Morton-Jones, D. *Polymer Processing.* Chapman and Hall, New York, NY, 1989.
44. Vane, L. and Rodriguez, F. Selective Dissolution: Multi-Solvent, Low Pressure Solution Process for Resource Recovery from Mixed Post-Consumer Plastics, *Proceedings of the Society of Plastics Engineers Regional Technical Conference — Recycling Technology of the '90's,* Chicago, IL, 1990.
45. Wielgolinski, L. A Family of Functionalized Acrylic polymers with Unique Solubility Properties for Recycling Applications, *Proceedings of the Society of Plastics Engineers RETEC Conference — New Developments in Plastics Recycling,* Charlotte, NC, 1989.
46. Bonsignore, P., Jody, B., and Daniels, E., Separation Techniques for Auto Shredder Residue, *Proceedings of the Society of Automotive Engineers International Congress and Exposition,* Detroit, MI, 1991.
47. U.S. General Accounting Office, Degradable Plastics-Standards, Research, and Development, Report to the Chairman, Committee in Governmental Affairs, U.S. Senate, GAO/RCED-88-208, 1988.
48. Leaversuch, R., Industry Weighs Need to Make Polymer Degradable, *Modern Plastics,* August 1987, p. 52.
49. Helmus, M. N., The Outlook for Degradable Plastics, *Spectrum,* Arthur D. Little Decision Resources, Boston, MA, February, 1988.
50. Budiansky, S., The World of Crumbling Plastics, *U.S. News and World Report,* November 24, 1986, p. 76.

CHAPTER 3

Practical and Legal Problems Relating to the Proper Disposal of Hazardous Household Chemicals/Waste

R Del Delumyea

INTRODUCTION

Approximately eleven tons of hazardous material are generated in Jacksonville daily. Most of this material ends up in the city landfill. Some people think that just because a substance can be purchased without restriction, it is not "really" harmful. Often, these products become a problem for the first time when they are ready for disposal. Information about the hazards of chemicals in the home, henceforth abbreviated HHW (Household Hazardous Waste), and proper methods for their disposal is available from a variety of sources. Library books are available on solid waste. The Water Pollution Control Federation had an excellent handout titled *Household Hazardous Waste. What you should and shouldn't do*; however, it is no longer in print. The Florida Department of Environmental Protection produces a booklet titled *Hazardous Waste from Homes*,[1] listing a number of common substances found in the home which have special disposal requirements. The U.S. Environmental Protection Agency's Office of Solid Waste provides, on request, publications on the subject. For information on programs in each state, the EPA has assembled a *Bibliography of Useful References and List of State Experts*.[2] The EPA offers a document to guide organizers of a HHW collection program at the community level.[3]

For the average homeowner, historical ways of disposing of these chemicals involved flushing them down the drain, pouring them on the ground, or placing them in the trash. For a number of commonly encountered chemicals, such practices are no longer acceptable. As a result, citizens, until recently, had no proper mechanism for the safe, environmentally benign disposal of household

1-56670-215-1/97/$0.00+$.50

hazardous chemicals. Most chemicals ended up in a local dump or landfill. There are probably hundreds of old (abandoned) private and public landfills/dumps in the City of Jacksonville. The locations of 128 dump sites were printed in the local paper. The city currently has one active landfill for all its household waste, including any hazardous materials which are improperly disposed in the trash.

In response to the recognized need to reduce the amount of these materials from the waste stream, the Solid Waste Disposal Division of the City of Jacksonville, FL, began sponsoring two annual collections of hazardous household chemicals from non-commercial residents in 1989, under the name "Amnesty Days." Specially designed collection sites were staffed by Solid Waste Disposal Division personnel. Individuals brought unwanted chemicals or commercial products for sorting and safe disposal. Materials collected were disposed of by a private waste disposal contractor hired by the City. The program was so successful that a permanent hazardous waste collection facility was established (a second site is planned). The facility is open each Tuesday (8:30–11:30 a.m. and 1:30–4:30 p.m.) and the second Saturday of each month.

STATEMENT OF THE PROBLEM

One requirement of the program is that the citizen must bring the waste to the collection point. Consequently, there are two segments of the population that have been unable to fully participate in the program, but which may represent significant sources of unwanted household chemicals — the elderly and the disabled. These people, particularly the former, may have accumulated a number of such chemicals over the years, with no way to safely dispose of them. Both groups may also be unable to handle and/or transport hazardous chemicals to the collection site. The author and Mr. Jerry Young, Department of Regulatory and Environmental Services, Hazardous Materials Control, City of Jacksonville, responded to two requests for assistance from elderly women whose husbands had died and left a garage/backyard with unknown chemicals (in one case 25 pounds of D.D.T.) To address this issue and to assess the magnitude of the problem, the Jacksonville Section of the American Chemical Society (ACS/JAX), working with the Solid Waste Disposal Division of the City of Jacksonville, proposed that qualified members of the Section go to the residences of people who sign up for the program, collect small quantities (100 kg) of household chemicals, and transport them safely to the collection site. The proposed work has been reviewed and discussed with the Solid Waste Division of the Department of Public Utilities of the City of Jacksonville, which supervises the actual collection site, and with the Hazardous Materials Control Division of Bio-Environmental Services.

The concept of the Jacksonville Section of the American Chemical Society acting on behalf of the elderly and/or disabled in the collection of household hazardous materials has received letters of support from: Mr. John K. Crum, Executive Director of the American Chemical Society; Ms. Kathy A. Hall, Director of the Retired Senior Volunteer Program of Duval County, which is part of the Human Services Department, Adult Services Division, of the City of Jacksonville; Mr. Jack Gillrup, Disabled Service Division, Department of Human Services, City of Jacksonville; Mr. Walter M. Lee, President of the Jacksonville Chamber of Commerce; and Mr. James O. Sewell, Associate Pollution Engineer for the Department of Regulatory and Environmental Services of the City of Jacksonville.

WHAT IS A HAZARDOUS SUBSTANCE?

Although there is a plethora of definitions for what is a hazardous substance, most break down the hazard to physical hazards (flammability, explosivity, corrosivity, or reactivity), health hazards (toxicity, eye, skin, or respiratory irritation) or those for whatever reason pose an environmental hazard (e.g., pesticides or heavy metals). Health hazards are either acute (immediate response) or chronic (cumulative response). A clear distinction exists between "hazardous" and "dangerous," although the distinction is often only valid to the definer. As a chemistry professor, the author often tells his students, "Chemicals are hazardous — Students are dangerous!" Various Federal agencies have definitions for what is hazardous. The type of hazard for a substance being discussed often indicates the nature of the agency responsible for its designation as hazardous. The Occupational Safety and Health Administration (OSHA) lists approximately 22 chemical hazards but there are essentially four common hazards: explosive, flammable, corrosive, or toxic (chronic or acute). Carcinogenic or cancer causing agents are another subcategory of health hazards.

A substance that is defined as hazardous but is commonly used, is gasoline. In addition to its flammability, gasoline contains aromatic hydrocarbons including benzene. Benzene is both an acute and a chronic hazard. Acute exposure to benzene can cause mucous membrane and/or respiratory irritation, restlessness, convulsions, excitement or depression and ultimately death through respiratory failure at high concentrations. Long-term exposure to low levels can result in bone marrow depression and aplasia, and (rarely) leukemia. Finally, benzene is considered to be carcinogenic. Nevertheless, as a petroleum product, its transport, storage, and use are not under direct control of the Environmental Protection Agency.

WHAT STATE AND FEDERAL LAWS APPLY TO HAZARDOUS WASTE DISPOSAL?

The Comprehensive Environmental Response, Liability and Compensation Act[4] (CERCLA) was set up in response to abandoned landfills, particularly those with hazardous waste. The concept of "cradle to grave" responsibility under CERCLA applies to anyone who has disposed of a hazardous substance in a leaking landfill. Any user may be a "Potentially Responsible Party" and subject to fines or cleanup costs, or both. Municipal landfills, and the household wastes disposed in them fall into a grey area of the law. The same chemicals which are considered hazardous when disposed of in large quantity, are (so far) not regulated when disposed of individually by a million people.

In 1976, Congress passed the Resource Conservation and Recovery Act (RCRA),[5] which charged the Environmental Protection Agency with defining what a hazardous waste is. According to this Federal law, household solid wastes are not considered as "hazardous waste," regardless of any potential hazard. This is specifically stated in the Federal Register:[6]

> (b) Solid Wastes which are not hazardous wastes. The following solid wastes are not hazardous wastes: (1) Household waste, including household waste that has been collected, transported, stored, treated, disposed, recovered (e.g., refuse-derived fuel), or reused. "Household waste" means any material (including garbage, trash and sanitary wastes in septic tanks) derived from households (including single and multiple residences, hotels and motels, bunkhouses, ranger stations, crew quarters, campgrounds, picnic grounds, and day-use recreation areas).

With this blanket statement, anything that goes in the trash is not considered hazardous by definition. Thus, it appears that collecting, packaging, and transporting household products will not be a violation of EPA regulations. The Department of Transportation uses the same definition. Florida Department of Transportation, which was informally contacted concerning their regulations, recommended a size limit of 100 kg of hazardous household chemicals per load.

Applicable state laws vary with the state; however, the state of Florida does not have an active household hazardous waste program. The city of Jacksonville does operate the above-mentioned program.

WHAT IS A HOUSEHOLD HAZARDOUS SUBSTANCE?

A typical citizen asked the question, "Do you have any neurotoxins in your home?" is likely to respond, "No!" — and be wrong. Household pesticides such as flea sprays and powders, "bug bombs," roach traps, and a variety of pest killers are neurotoxins. Similarly questioned about explosives, he/she may not give a different response, yet fertilizers (mainly ammonium nitrate) are potentially explosive. Spray cans, such as hair sprays are potentially explosive due to their internal pressure, and the use of volatile, flammable gases as propellants. Aerosol grade propellants, as sold by Phillips Petroleum, are mixtures of propane, butane and isobutane, in proportions according to the volatility required.[7] Hydrocarbon Propellant A-31®, 95% isobutane is also the fuel used in butane lighters (e.g., Bic®). Spray paints likewise use volatile, flammable hydrocarbons as solvents. The restrictions on the use of Freons,® which are non-flammable, have caused solvent propellant use to become greater.

Personal care products can also be hazardous. Solvents in cleaning fluids may be chemical hazards by the nature of the chemical used; nail polish remover (acetone) and the nail polish itself, hair sprays, and other widely used products are flammable, the former being under pressure as well.

Drain cleaners and oven cleaners, which contain sodium hydroxide (a strong caustic), can cause skin burns. Other cleaning products such as window cleaners contain ammonia, which can cause skin and eye irritation. Floor and furniture polishes contain various hazardous ingredients such as diethylene glycol, petroleum distillates or nitrobenzene. Pool chemicals such as pH adjusters contain acids. Pool chlorine, either solid or liquid, contains hypochlorite, which can release gaseous chlorine (a powerful oxidizer and strong respiratory and eye irritant) under certain conditions. Similarly, laundry bleaches often contain hypochlorite. The brick cleaner, muriatic acid, is hydrochloric acid, a corrosive and an eye, skin and respiratory irritant. Mothballs are either naphthalenes (a poison) or paradichlorobenzene (eye, skin, and respiratory irritant), are both toxic by inhalation and skin absorption at high concentrations.

Photographic chemicals contain such hazardous chemicals as silver, acetic acid, hydroquinone, sodium sulfite and ferrocyanide, many of which are environmental, if not health, hazards. Art supplies, particularly brightly colored oil or acrylic paints contain metal compounds which are hazardous. The name of the color often indicates the ion of concern: cadmium yellow, lead white, manganese blue, molybdate orange, strontium yellow, or zinc yellow.

The chemicals found in a typical garage can contain a variety of hazardous chemicals: antifreeze is ethylene glycol (toxic); transmission fluid (hazardous, toxic); brake fluids contain glycol ethers and, possibly, heavy metals (flammable, toxic); used engine oils contain hydrocarbons and heavy metals (flammable, toxic); and old lead-storage batteries contain sulfuric acid (corrosive) as well as lead metal, which can be recycled. Oil-based paints, until dry, contain solvents which can be flammable or toxic. And finally, the most accepted hazardous materials are lawn and garden products: fertilizers, pesticides, insecticides, fungicides, and rodenticides.

THE JACKSONVILLE PICK-UP PROGRAM

Based on the above experiences, the author, then the Chair of the Jacksonville Section of the American Chemical Society (ACS/JAX), worked with local individuals to try to secure funding for the project. ACS/JAX includes persons trained in the proper handling of hazardous materials through their jobs with local transportation, chemical, industrial, and educational professions. ACS/JAX proposed to provide personnel and equipment to transport hazardous household chemicals to the Amnesty Site designated by the Solid Waste Division of the Department of Public Utilities of the City of Jacksonville for persons who meet certain eligibility requirements. The initial scope of the work to be performed is in the nature of a pilot study. Perhaps a reflection of the size of the project and the quantity of materials in each load, it has been nicknamed the "Pick-up Program." It is intended to identify the problems (and solutions to them) which might be encountered in the full-scale collection of household hazardous materials for disposal. Upon completion of the pilot study, the success will be evaluated and a determination will be made whether to expand the program at the local level or seek external funding. A larger program would likely require state or federal funds. The following is a description of the program as currently envisioned. Funding has not yet been secured.

Identification of Eligible Participants

Persons eligible for the Pick-up Program must meet the following description: "Physically confined to the home and unable to bring the material to the disposal site." It is anticipated that the majority of the participants will be elderly persons, with a smaller number of individuals with disabilities.

Awareness of the Program

To make those segments of the population aware of the program, the assistance of local agencies, of which there are approximately 50 in the Northeast Florida area, will be required. Contact has been established with such agencies as: the City of Jacksonville Disability Services; Retired Senior Volunteer Program; City of Jacksonville Human Services, Senior Social Service Project; and the Area Agency on Aging (part of the Florida Council on Aging). Several agencies have agreed to running announcements of the program in their newsletters. Flyers and public service announcements on local radio and television will announce the program. The recent emphasis by the ACS on Public Outreach, including training for the Section's Public Outreach and Media Relations Chair, will further enhance the ability to publicize the program. A brochure describing the program will be prepared and distributed in advance of each Amnesty Day.

The City advertises the Amnesty Days prior to each collection period. Funds will be sought to assist the advertisement of the Amnesty Day program. The program will be described in advertisements in the Florida Times-Union along with announcements of the Amnesty Days. A telephone number will be announced where those wishing to have items picked up can call for a site visit.

ACS/JAX has been in contact with representatives of the City of Jacksonville's Department of Human Services, Senior Social Service Project; the Florida Council on Aging (Jacksonville Chapter); the Retired Senior Volunteer Program; and has presented the proposed work to the Mayor's Disability Council (which provided a letter of support), a group composed of individuals and organizations representing individuals with disabilities, which addresses issues facing the disabled and makes recommendations to the Mayor of the City of Jacksonville. The Jacksonville Chamber of Commerce has similarly endorsed the project and provided a letter of support. Responses to the proposed collection program from all concerned have been enthusiastic and encouraging. Numerous offers of assistance in identifying, contacting, and assisting the elderly and disabled community have been received.

Types of Chemicals to be Collected

The Pick-up Program will collect almost any items, provided they are in containers in good condition and have a clearly readable label. In order to determine the acceptability of any particular item(s), a site visit will be made prior to the date of the actual collection. At that time, the resident will be asked to fill out a form allowing people on their property and to sign a liability waiver. A list of chemicals which are to be taken on the pick-up day will be prepared and checked again when the vehicle arrives to remove the substance(s). Each type of hazard will be boxed separately but multiple sites may be visited before the shipping container is full. With adequate planning, and depending on the size of the collection planned, multiple trips may be possible during which only one class of chemicals may be transported.

Safety Equipment Required

Each vehicle will be equipped with boots, gloves, and splash aprons for each person and a respirator in the event of an accidental spill during packing. It is anticipated that the volunteer will be able to secure the loan of the necessary equipment from his/her work-place. The types of chemicals which will be collected involve no more hazard than would be found in the average house or garage. In the event that a person determines an area to be hazardous, he/she will be instructed not to enter and to notify the resident of the situation, and to take no further action.

Transportation of Materials to Disposal Site

One of the first questions raised involved the perception that ACS/JAX was becoming a hazardous waste transporter. If the waste were being transported by its owner, it is exempt from regulation. Since it is taken by someone other than the owner, its ownership may be questionable. Regardless, household wastes are not considered hazardous by some agencies and the project was looked at with suspicion by some who feared the volume or quantity of compounds involved might be too high. Correct packing and segregation of wastes by hazard minimize the chance of an accident. After discussions with Florida Department of Transportation representatives, it was concluded that, although such wastes are probably exempt, restricting the loads to small quantities (100 kg) of household chemicals should avoid any problems.

Containers will be transported in cardboard containers and surrounded by absorbent material. The materials will be separated by hazard category. A maximum of 100 kg (approximately 200 lbs) of material will be transported at a time.

Number of Vehicles Involved

For the first-time pilot project, two pick-up trucks will be used — one on the west side of the St. Johns River, one on the east side. These vehicles must be supplied by members of the ACS-Jacksonville Section, as city and state vehicles are prohibited from involvement in the collection and transport process.

Qualifications of Individuals Involved in Handling Materials

Only volunteer members of the Jacksonville Section of the American Chemical Society who are knowledgeable in the safe handling of chemicals will be involved. At least one of the persons on each vehicle will have completed an OSHA-approved "40-Hour Hazardous Substance Health and Safety Training" course. Such courses are offered regularly at the University of North Florida Division of Continuing Education and the Environmental Education and Safety Initially, ACS National was contacted for support of the project. Although eagerly supported in theory, the Society determined that it would be impossible to financially support this single project since, given its appeal and likely success, every ACS section might want to participate. Subsequently, a grant request was submitted to the Jacksonville Rotary Club. Although reviewed favorably, it was not funded.

The cost of conducting periodic pick-ups is relatively small. Participants are volunteers. The Solid Waste Division of the City will provide containers for transporting the waste and pay for its ultimate disposal. Local industries have volunteered to provide most miscellaneous items such as packing material and boxes for transport. Gasoline for the vehicles, and meals for the volunteers on the day of collection will be taken care of by ACS/JAX. Personal safety equipment (e.g., respirators, safety glasses, boots, etc.) will be provided by each participant, since these individuals are involved in similar work professions. Additional items, if required, (such as over-packing containers) will be obtained by loan or donation from companies or institutions whose employees are members of the ACS.

Use of vehicles from a local (Jacksonville University) school was ruled out (for insurance reasons). The University, a Small Quantity Waste Generator, employs a private firm to handle the disposal of its hazardous waste. That company was contacted but did not volunteer to assist the project. It did provide the name and contact for its insurance company, which was asked what it would cost to insure two vehicles for two days twice each year. It was not interested. A company in Alabama was found which would write a liability policy with a one million dollar limit, suitable for two vehicles, two weekends during one year. The cost of that policy was $25,000! Further, the company would only make the offer if the two vehicles were covered on a separate policy for the rest of the year, making the additional coverage essentially a rider on the conventional insurance policy.

Local waste haulers (e.g., BFI Industries) were contacted as possible sponsors of the Pick-Up Program. Although agreeing that the materials often end up in their trucks, none has been willing so far to enter into the potential mine field of liability.

INSURANCE AND LIABILITY

The EPA exempts household chemicals from restrictions on transportation, storage or disposal. It must be remembered that these are chemicals which can be purchased without restriction, which can be driven home, taken on a bus, or in a disabled person's vehicle. At a recent meeting with City of Jacksonville staffers, it was pointed out that such materials are being transported already — on city buses, in COMSYS (disabled services) vans and, significantly, garbage trucks. During the summer of 1992, a fire in a garbage truck in a residential area resulted in the entire load being dumped in the street to get at the burning material. The cause of the fire was not determined; however, it could have resulted from disposal of incompatible materials, an oxidizer, or other chemical reaction. A mechanism to remove these materials from the waste stream is desired; however, as with any potentially dangerous task, the question arises as to who pays in case of an accident. Without doubt, the single thing which has prevented the project from beginning is the need for liability coverage in case of an accidental spill of a compound.

The American Chemical Society owns and operates vehicles in the Washington area. When the issue of insurance came up, it was thought that perhaps the insurance company which provides coverage for ACS vehicles would be interested in adding a rider to the policy. Humorous as it is, the insurance policy covering the American Chemical Society's vehicles has a lengthy exemption covering chemicals in a one-and-a-half page EXCLUSION — ALL POLLUTION INJURY OR DAMAGE policy endorsement. Any damage, loss, or injury caused by chemicals carried in ACS vehicles is not covered. Period. This turned out to be the case for all automobile insurance policies. Jacksonville calls itself the "Insurance Capitol of the South," being home to (among others) American Heritage, Blue Cross/Blue Shield, Peninsular, Prudential, and State Farm Insurance. It was explained that automobile insurance companies are not in the "liability business." Unfortunately, the City of Jacksonville is prohibited by statute from providing city-owned vehicles for the project.

In order to release the ACS/JAX from liability for any problem that might arise during pick-up and packing the chemicals, a Disclaimer similar to the one used for ACS inspections of laboratories was developed. As is always the case, such a disclaimer is only as strong as the lawyer defending it. Medical coverage of the volunteers during the project will be up to the individual volunteering for the work.

ACKNOWLEDGMENTS

Mr. Jerry Young, Associate Pollution Control Engineer in the Hazardous Materials Control section of the Department of Health, Welfare and Bio-Environmental Services of the City of Jacksonville also deserves to share credit for the concept of the ACS Pick-Up Program. Young was the person who responded to the incident involving an elderly widow and a bag left/found in her basement after her husband's death. The bag contained 25 pounds of D.D.T., a pesticide banned for several years. Young is acknowledged for both encouraging and assisting the author in interactions with the city.

Mr. Steve Waterman, Solid Waste Special Assistant with the Department of Public Utilities of the City of Jacksonville, currently runs the Household Hazardous Waste collection facility. He was involved in the original planning of the ACS Pick-Up Program. Waterman sometimes requires the assistance of the author in handling unusual requests for his assistance that may require additional chemical information available to a Chemistry Professor.

REFERENCES

1. *Hazardous Waste from Homes*. Florida Department of Environmental Education, Tallahassee, FL, 1988.
2. *Household Hazardous Waste. Bibliography of Useful References and List of State Experts*. EPA530-SW-88-014. USEPA, Office of Solid Waste (OS-305), Washington, D.C., 1992.
3. *Household Hazardous Waste Management. A Manual for One-Day Community Collection Programs.* EPA530-R-93-026. USEPA, Office of Solid Waste (OS-305), Washington, D.C., 1993.
4. Comprehensive Environmental Response, Liability and Compensation Act of 1980. U.S. Code 42, Ch 9601. Public Law 96-510, U.S. Government Printing Office, Washington, D.C., December 11, 1980.
5. Resource Conservation and Recovery Act of 1976. U.S. Code 42, Ch. 6901. Public Law 94-580, U.S. Government Printing Office, Washington, D.C., October 21, 1976.
6. Code of Federal Regulations, Vol. 40, Part 261.4(b)(1). Identification and Listing of Hazardous Waste, 1995.
7. *Hydrocarbon Propellants: The Pure Solution,* Phillips 66 Company, Roanoke Rapids, NC, 1986.

CHAPTER 4

Household Hazardous Material: The Evolution of State Initiatives

Joan D. Sulzberg and Richard K. White

INTRODUCTION

Products and materials stored and used in the home contain a broad range of chemicals and are integral in the daily life of Americans. The average American home contains 63 chemical products which may be classified as hazardous.[1] Items such as cleaners, paint products, yard maintenance products, automotive products and household batteries contain substances that are considered hazardous by federal legislation, when generated in large quantities. Of the one to two million calls received each year, at Poison Control Centers nationwide, since 1986, an average of 92% are related to exposures in the home.[2-8] Regarding specific materials, an average of 10% of the calls are related to cleaning substances, 8% related to cosmetics and personal care products and 3.7% related to pesticides. Although these products contain constituents which are regulated by federal hazardous waste regulation, they are exempt when generated for use in the home. As a result of this exemption and the controversial nature of household hazardous material (HHM), communities and states nationwide have developed unique methods for managing these hazardous substances.

This chapter presents background on HHM by (1) putting HHM in context with respect to historical hazardous waste management legislation and practices, (2) discussing specific definitions and estimated amounts of HHM in the municipal solid waste stream, and (3) describing the risks and pathways for exposure from HHM. From this background, the rationale and goals of selected state level HHM initiatives will be explored, followed by an assessment which will include

1-56670-215-1/97/$0.00+$.50
© 1997 by CRC Press, Inc.

examples of programs in the three phases of a HHM: pre-consumer, consumer, and post-consumer. Finally, implications of the research will be examined.*

PUTTING HOUSEHOLD HAZARDOUS MATERIAL IN CONTEXT

There is no clear-cut definition for products and materials used in the home that contain hazardous substances and several terms exist that describe such materials. *Household hazardous products, household hazardous substances, household hazardous materials, household hazardous wastes,* and *toxic household substances* are several terms that are commonly used. *Household hazardous waste* is, by far, the most widespread term. Since each term has a certain connotation associated with it, this chapter will use the term *household hazardous materials* to reflect the all encompassing nature of the issue. That is, HHM refers to products manufactured by industry; products sold in retail establishments; products used in the home and residuals of products that are to be disposed of.

Household Hazardous Material in Relation to Hazardous Waste

Although HHM have been used in homes for years, in the past two decades there has been a move to control the manufacture, use and disposal of such materials. To understand why this has occurred, one must look briefly at the history of hazardous waste management in the U.S.

In 1976, Congress passed the Resource Conservation and Recovery Act (RCRA) as it revised and amended the 1965 Solid Waste Disposal Act and the 1970 Resource Recovery Act. RCRA directed the U.S. Environmental Protection Agency (EPA) to develop and implement a program to protect human health and the environment from the effects of improper solid and hazardous waste management practices. Unique in its management of hazardous waste, the program was designed to control the management of such wastes from initial generation to ultimate disposal — from cradle to grave.[9] Initially, the focus of RCRA was aimed at large companies which generated a major portion of the hazardous waste in the U.S. In 1980, smaller companies, defined as those generating less than 1000 kilograms (2,200 pounds) of hazardous waste in one month, were exempted from the hazardous waste regulations to which larger businesses had to adhere.[9]

In the 1984 Hazardous and Solid Waste Amendments to RCRA, Congress directed the EPA to lower the small business standard to 100 kilograms (220 pounds) per month. In 1986, the EPA promulgated revised regulations that lowered the generation rate limits. Small businesses that generated between 100 kilograms and 1,000 kilograms became known as small quantity generators and

---

* To narrow the scope of this chapter, only state level initiatives are discussed, although many cities and counties have developed policies and programs for the management of HHM. The research presented here is part of a larger effort in which California, Florida, Iowa, Minnesota, New Jersey, New York, Pennsylvania, and Washington were targeted for an in-depth study. The authors have undertaken the reported study in order to assist the state of South Carolina in developing a comprehensive policy and program for handling household hazardous waste.

had to comply with different rules than the larger companies. Businesses that generated less than 100 kilograms per month of hazardous waste continued to be exempt from most hazardous waste requirements and became known as conditionally exempt small quantity generators. Interestingly, approximately 90% of the nation's hazardous waste was produced by large quantity generators, but these generators constitute approximately 2% of the companies covered by RCRA.[9] The remaining 98% of the businesses were classified as small quantity generators.

Since the enactment of RCRA and its amendments, federal and state environmental agencies have demonstrated success with controlling hazardous waste from easily identifiable point sources of pollution, such as large industrial outputs. At the same time, harder to identify, non-point sources of pollution, such as agricultural runoff, storm water runoff, conditionally exempt small quantity generator waste, and hazardous materials from homes, were passed over. It is only recently that these non-point sources of pollution have been identified and targeted for both pollution prevention and pollution control mechanisms. As a result, each state was left to take initiative in managing the less prevalent hazardous-containing wastes, such as HHM.

Defining Household Hazardous Material

As part of RCRA, a number of wastes were specifically excluded from being considered hazardous. RCRA Subtitle C, Section 261.4(b)(1) unconditionally exempts household wastes from being designated as hazardous, even when accumulated in quantities that would otherwise be regulated or when transported, stored, treated, disposed, recovered, or reused. The EPA interpreted this exclusion to include materials that

1. were generated by individuals on the premises of a temporary or permanent residence and
2. were composed primarily of materials found in the wastes generated by consumers in their homes, even if the waste exhibited one of the four characteristics of hazardous waste — ignitability, corrosivity, reactivity, or toxicity.[10]

A general definition of *household hazardous material* is any material containing a substance that is regulated by the EPA under RCRA or any material that fails one of the characteristic hazardous tests, is stored at home, and is typically discarded in the municipal solid waste stream.[11a,12] State definitions are comparable to this definition.

For instance, California Health and Safety Code Chapter 6.5, Article 10.8, Section 25218.1(e) defines that, "Household hazardous waste means any hazardous waste generated incidental to owning and/or maintaining a place of residence. Household hazardous waste does not include any waste generated in the course of operating a business concern at a residence." The definition of hazardous waste in California is more stringent than the federal definition and thus includes more substances and materials. In Florida, the interpretation of HHM is "nonregulated waste that exhibits one or more of the characteristics of hazardous waste that is generated by a household or other non business sources."[13] Iowa opted to characterize HHM in a distinctive way. The Iowa definition includes products used for residential purposes that are hazardous by Iowa Code, but excludes materials such as laundry detergent or soaps, dishwashing compounds, chlorine bleach, personal care products, personal care soaps, cosmetics and medication.[14]

The composition of HHM may vary by community, by season, by degree of urbanization and by socioeconomic class. Rathje and Murphy[15] found that a large proportion of HHM from low income homes were automotive products; a significant amount of HHM from middle income homes were home improvement products such as paints, stains, and varnishes; and the HHM of affluent homes contained more lawn and garden related products. Another phenomenon noted by Rathje and Murphy[15] is that contrary to what was intended, the garbage discarded after a well publicized HHM collection day contained more than twice as much hazardous material, by weight, than the garbage that had been discarded before the collection day. Since residents became aware of the hazards of the material stored in the home but missed the collection day opportunity, they decided to rid themselves and their homes of the hazardous material in the conventional manner, in the garbage. A sample breakdown of HHM found in the municipal solid waste stream, by weight, is shown in Table 4.1.

Determining the Amount of Household Hazardous Material

One method used to determine the amount of HHM in the municipal solid waste stream is a waste composition study or garbage sort. This entails a physical separation of waste stream components to calculate the percentages of each. A solid waste composition study conducted by the Minnesota Pollution Control Agency in 1990–1991 found that from 0.6% to 1.2% of the solid waste stream was hazardous, including oil filters, with the average being 1.0%.[16] The Garbage Project of the University of Arizona has conducted studies in garbage archaeology since the early 1970s. After performing garbage sorts and collecting HHM infor-

Table 4.1 Sample Categories and Percentages of Household Hazardous Material[67]

Household hazardous material category	Percentage, by weight
Home maintenance	37
Household batteries	19
Personal care	12
Household cleaners	12
Automobile maintenance[a]	11
Yard maintenance	4
Other	5

[a] Does not include lead acid batteries.

From Office of Technology Assessment, *Facing America's Trash: What Next for Municipal Solid Waste?*, Government Printing Office, Washington, D.C., 1989.

mation, Rathje and Murphy,[15] of the Garbage Project, contend that about 1% of all household garbage, by weight, was hazardous. In the Los Angeles area, it was estimated that 0.7% of all residential waste was hazardous.[17] Earlier research, discussed by Kinman and Nutini,[18] found HHM percentages to be small, sometimes less than 0.01%. Duxbury[19] of the Waste Watch Center, maintains that 0.5% is the national average.

Despite the differences in the projected amounts of HHM found in the municipal solid waste stream, there is no question of its existence. In 1990, 196 million tons of municipal solid waste were discarded in the U.S., and by the year 2000, this number is projected to reach 216 million tons.[11] Given an estimate for HHM in the solid waste stream as 1%, when multiplied by the amount of municipal solid waste generated in the U.S., HHM totals 1.96 million tons in 1990 and a projected 2.16 million tons by 2000.

There are two flaws in determining the amount of HHM in the manner described above. First, although waste sorts are an effective method of determining hazardous constituents in the municipal solid waste stream, they do not include any of the materials that are stored in the home, or discarded through other avenues, such as the sewer, sink, backyard or home burning.[20] Haugh et al.[21] report that the average rural home in Washington State generates 27 pounds of HHM per year while the average urban household generates 23.6 pounds per year. The second issue of concern relates to the recent prevalence of non-hazardous material recycling efforts. Since materials such as paper, plastic, and aluminum are being diverted from the solid waste stream, by up to 50% in some states, the percentage of hazardous material in the municipal solid waste, by simple mathematics, would be expected to increase.

Risks and Pathways of Household Hazardous Material

HHM present risks to human health and the environment. Since components of HHM do not lose their hazardous characteristics when discarded into the municipal solid waste stream,[22] they are simply transferred to another location or medium. Although it seems logical to assume that HHM contribute to household accidents, such as fires and injuries, there is a lack of a national data to support this. Firefighters have expressed concern regarding the impacts of fighting chemical fires in a household setting. The situation in which a firefighter encounters an exploding aerosol can or inhales burning pesticides or other toxic vapors contributes to these concerns.[20] An increased risk of accidental poisoning occurs when HHM accumulate. As cleaners, pesticides, and other HHM are often stored in easy access locations, children are exposed to poisonous and harmful agents by default. Of all the calls received by Poison Control Centers between 1986 and 1992, an average of 46% were related to children under the age of three and 61% related to children under the age of six.[2-8]

In the home, improper mixing of certain HHM can facilitate violent reactivity. Chlorine bleach mixed with an ammonia based cleaning product, for instance, will react to form chloramine, a deadly gas. Likewise, indoor air pollution is an indirect effect of home storage and usage of certain HHM. Accordingly, the EPA has identified indoor air pollution as a major pathway of human exposure to numerous chemicals and is working on the development of a system to review a large number of potential indoor air sources and assign priorities for further evaluation.[23]

The storage and transport of HHM with municipal solid waste can lead to spills and accidents as well. Mitchell and DeMichelis[24] reported situations where items such as swimming pool chemicals, solvents, paints, acids, and aerosol cans have caused injury to sanitation workers when the wastes have spilled or splashed onto the worker. A California study in 1982 found that 3% of all refuse collection workers in the state were injured due to contact with HHM mixed in with trash and garbage.[25] In Minnesota, in 1982, a small container of an unknown flammable solvent broke open in the shredder of a refuse-derived fuel incinerator, vaporized and ignited. The resulting explosion caused equipment damage totaling $800,000 and consequently forced a two-year shutdown for repair.[20]

Illegal dumping of HHM can lead to the flow of hazardous chemicals and liquids into groundwater and surface water supplies which many cities draw on for drinking water. One gallon of illegally dumped motor oil has the potential to contaminate up to one million gallons of water. This may become a large scale problem because do-it-yourself oil changers improperly dispose of greater than 200 million gallons of used oil nationwide each year. Infiltration of oil into ground and surface waters may also lead to the contamination of soil, fish, and wildlife.

Contamination of groundwater at and surrounding sanitary municipal solid waste landfills with materials such as organics and heavy metals has been documented, even at sites that are distant from industrial sources.[27] Ferry[28] reports other instances where groundwater contamination from organics has occurred at municipal solid waste landfills in states such as Oklahoma, California, New Hampshire, Wisconsin, and New York.

Due to the haphazard nature of traditional waste disposal methods, the Office of Technology Assessment[29] estimated that approximately 22% of the 850 sites proposed for the original Superfund National Priority List were municipal landfills. Industrial wastes were considered the most significant source of contamination at these sites, followed by sewage sludge and HHM. A Connecticut court case decided a municipality can be held liable for cleanup costs at a Superfund site as a result of its contribution of solid waste.[30] Although RCRA exempts HHM from industrial hazardous waste regulations, Superfund does not. This decision was affirmed by the 2nd Circuit Court of Appeals.[30,31] "This means that if household trash trucked to landfills or other contaminated sites can be shown to have hazardous constituents, local governments and, therefore, local taxpayers must share cleanup costs."[31]

Pollution problems arise in gas emissions and ash from burning HHM in combustion facilities (and illegally in the back yard). Major concerns involve heavy metals, such as mercury, cadmium and lead. Household batteries and automotive batteries were identified as the dominant sources of mercury and lead, respectively, in incinerator ash.[32,33]

Finally, households and commercial establishments can discharge significant amounts of toxic and other harmful pollutants into domestic wastewaters.[34] Galvin[25] explained that septic tanks, drain fields, and municipal sewage systems are all susceptible to the hazards of HHM, especially heavy metals and organic loads. Concerning sewage systems, sewer pipes may corrode and sewage effluent and sludge may be contaminated from HHM being poured down the drain. Primary and secondary treatments of wastewaters were not designed to remove the chemicals contributed to the waste stream by HHM.[25] As a result, effluent into receiving waters may contain contaminants. Also, sewage sludge is used on cropland. It is therefore necessary to minimize the contaminant load.

STATE HOUSEHOLD HAZARDOUS MATERIAL INITIATIVES

Due to the exempt status of HHM from federal hazardous waste regulations, the risks associated with the hazardous materials and increasing citizen awareness and public pressure, individual states have developed state-wide management programs. The development of a HHM plan for Florida stemmed from the state's dependence on groundwater and its susceptibility to groundwater pollution.[35] Prior to implementation of hazardous waste reduction and cleanup efforts, approximately 1,000 hazardous waste sites were identified in the state.[36,37] Also, it was determined that the Biscayne Aquifer, a major drinking water source in Florida, was contaminated. Since over 90% of Florida's residents depend on groundwater for drinking,[38,39] it was deemed necessary to protect the integrity of the valued resource. The Water Quality Assurance Act,[40] passed in 1983, established the foundation of a comprehensive hazardous waste management program and provided for the Water Quality Assurance Trust Fund and Amnesty Days — a one time opportunity for residents to dispose of accumulated HHM.

Iowa has also experienced groundwater contamination. With 75% of the ground and surface water potentially contaminated, and 82% of the population depending on groundwater for drinking water, it was apparent to the Iowa legislature that action had to be taken.[37] A permanent HHM program was established as a result of the 1987 Iowa Ground Water Protection Act. The purpose of the program was to collect and dispose of small amounts of hazardous waste being stored in residences, or on farms in a particular county.[41,42] In addition, education of the public with respect to source reduction and proper use, storage and disposal of HHM were goals of the program.[42]

A similar rationale was evident in Minnesota:[43]

Minnesota has a total of 133 landfills, of which 53 are operating. Of the 133 landfills, 62 are included on Minnesota's State Superfund list for cleanup of contaminated groundwater. Furthermore, 11 of the 62 landfills on the State Superfund list are also included on the list of Federal Superfund sites. Since almost half of the total amount of landfills in Minnesota require some type of cleanup actions, reducing the amount of household hazardous waste entering landfills helps reduce groundwater contamination at remaining operational landfills. This will reduce landfill cleanup activities and costs.

Although the focus was on groundwater and landfill issues, the reduction of future liabilities and costs was also a main component of the rationale. In addition, Goldsmith,[44] of the Minnesota Pollution Control Agency (MPCA), cited a 1982 explosion (discussed earlier) as a concern. Gilkeson[45] of the MPCA reported that HHM issues were brought to the forefront of the state agenda along with the incineration boom of municipal solid waste in Minnesota in the 1980s because heavy metals in air emissions were a concern.

The California legislature found:[46]

1. That because hazardous substances are an integral part of daily life, it would benefit the public to have access to practical and consistent information concerning chemicals in daily life, products which contain hazardous substances, and proper procedures for the disposal of hazardous substances. This information would improve the ability of all Californians to assist in protecting the state's natural resources from environmental degradation.
2. The disposal of hazardous substances by households can be injurious to sanitation workers, the general public, and wildlife and domestic animals, and can pose a threat to the environment.
3. Each household in the state should have reasonable access to legal, convenient, and environmentally safe methods for the disposal of hazardous substances commonly found in and around homes.

Another reason for the California devotion to the legislation and management of HHM was a result of the landfill crisis.[47] Californians generate more municipal solid waste, per capita, than every other state except Maryland.[1] The thrust of state legislation advocated the diversion of as much solid waste as possible from municipal landfills, including HHM. Additional reasoning behind HHM programs in California included potential municipal liability issues.[48]

The Washington state goal regarding HHM, was to divert HHM from RCRA Subtitle D (solid waste) disposal facilities to Subtitle C (hazardous waste) disposal facilities.[49] In New Jersey, the thrust of the HHM management scheme was to develop an environmentally sound, cost effective program to remove constituents that prevent the reuse of municipal solid waste. The goal was to "clean out" the solid waste stream to support projects such as landfill mining.[50]

LIFE CYCLE ASSESSMENT OF STATE INITIATIVES

It is now appropriate to explore specific state level HHM initiatives aimed at reducing damage to human health and the environment. The general life cycle of HHM can be broken down into three phases: the pre-consumer phase, the con-

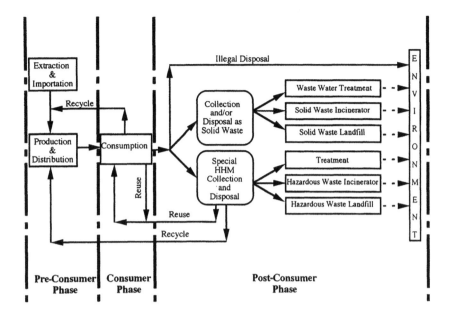

Figure 4.1 Life cycle phases for HHM.

sumer phase, and the post-consumer phase. The movement of hazardous material through these phases is depicted schematically in Figure 4.1. As shown in the diagram, materials move from the extraction, importation, production and distribution phases, to the consumer phase, and then either to reuse, recycle, or disposal. The following sections describe specific practices of several states relating to these life cycle phases. The states discussed are California, Florida, Iowa, Minnesota, and Washington primarily because these states exhibited a diversity of comprehensive and mature HHM practices and personnel at the state environmental agencies were cooperative by willing to participating in the study. The programs discussed are not comprehensive for the particular state; rather, they are indicative of the types of programs that are occurring in the HHM arena.

Pre-Consumer Phase

This phase includes programs and initiatives targeted at the manufacturer, the wholesaler or importer, as well as the retailer of HHM. Below are several examples of programs targeted at the pre-consumer phase.

Example 1. Manufacturing: Household Batteries

The first state to pass comprehensive household battery legislation was Minnesota. Legislation required manufacturers to reformulate products and take responsibility for collecting and recycling spent products. With respect to mercury content, the following rules were set:

1. mercury in alkaline batteries were limited to 0.30%, by weight, by 1991 and 0.025% by 1992,
2. mercury per button cell, non-rechargeable batteries were limited to 25 mg and,
3. sale and distribution of dry cell batteries containing a mercuric oxide electrode were banned.[51]

Also, the legislation required new battery types to be approved by the MPCA Commissioner. With respect to rechargeable batteries and items, specifications were set that prohibited the sale of any product containing rechargeable batteries unless:

1. the batteries were removable or in a battery pack,
2. both the battery and product were labeled with contents, and
3. the package was labeled with instructions addressing battery recycling and disposal.

To continue sales in Minnesota, manufacturers of household rechargeable batteries were also required to conduct collection programs.

Example 2. Manufacturing: Volatile Organic Compounds

California has long regulated the release of ozone-forming compounds from cars and factories, but effective 1992, the California Air Resources Board adopted laws to substantially reduce volatile organic compounds from several consumer products.[52] Standards for the percentage of volatile organic compounds, by weight, in each of 12 products, were mandated by January 1993, and for 16 additional products by January 1994. Lower percentages were then set for five products by 1996, 1997, and 1998. Products targeted for reduction include air fresheners, automotive windshield washer fluids, bathroom and tile cleaners, engine degreasers, floor polishes and waxes, furniture maintenance products, general purpose cleaners, glass cleaners, hair sprays, hair mousses, hair styling gels, laundry prewash, nail polish remover, oven cleaners, insect repellents, and shaving creams. The Air Resources Board then adopted regulations that added products such as aerosol cooking sprays, automotive brake cleaners, charcoal lighter material, carburetor-choke cleaners, dusting aids, fabric protectants, household adhesives, insecticides, laundry starch products, and personal fragrance products. Additionally, antiperspirant and deodorant regulation specified standards for volatile organic compounds in aerosol deodorant and antiperspirant products, including product forms such as aerosols, pumps, liquids, and solids.[53] For example, medium volatility organic compounds, by weight, are limited to 10% in aerosol antiperspirants, effective January, 1995.

Example 3. Retailer: Required Permits and Shelf Labeling

In Iowa, management of HHM requires every retailer who sells products identified as household hazardous material to obtain a $25 annual permit.[41] As of January 1993, there were approximately 12,800 permitted HHM retailers.[42] To ensure that retailers obtain the necessary permits, the Iowa Department of Natural Resources (IDNR) worked with the revenue agency who sells the permits and established a computer program to monitor sales tax permits and HHM permits by Standard Industrial Classification (SIC) Code. Retailers in SIC codes that have a high probability of selling HHM (such as hardware stores) but do not have a HHM permit are sent a letter alerting them to the law and requiring them to either obtain a permit or sign an affidavit that they sell no HHM. After a year of passage, compliance of 30–40% was estimated by the IDNR.[54]

Retailers of HHM in Iowa were also required to label shelves beneath HHM with yellow circular stickers that signify a product is hazardous. The label depicts a container of oil being dumped onto the state silhouette with a line across it. Retailers must also place posters and consumer information booklets in locations near to areas that contain large quantities of hazardous products.[55] The monitoring of the retail establishments for proper display of HHM decals and information packets has been accomplished through a joint effort of county sanitarians, environmental activists, and other interested parties. Compliance and enforcement have been poor.

Example 4. Financial Mechanism: Hazardous Substance Tax

In 1988, the Washington Model Toxics Control Act was passed by a vote of the people (citizen initiative) by 70%. This Act authorized the establishment of a hazardous substance tax on the first possessor of a hazardous substance in Washington State. The tax was set at 0.7% of the wholesale value and generates about $40 million each year.[49,56] More than 85% of receipts from the tax come from petroleum products. Of the total amount collected from the hazardous substance tax, 53%, or $21 million, goes into the Local Toxics Control Account. Of the $21 million, one-half goes to cleanup activities and one-half to preventive programs, such as hazardous waste planning, HHM education and information, solid waste planning and enforcement, waste reduction and recycling, groundwater monitoring, and landfill closures.[57]

Consumer Phase

This phase predominantly attempts to increase consumer awareness and alter purchasing and use habits. The following examples provide an indication of varying state efforts to accomplish this objective.

Example 1. Education Effort: Multi Media Approach

The MPCA believed that education of Minnesota citizens to reduce the generation of HHM would result in a direct cost savings from avoided disposal. As such, education is a primary focus of the permanent HHM effort. One goal of the Minnesota HHM program is for counties to become as self-sufficient as possible with respect to education and finance.[58] To assist counties, the MPCA developed a modifiable model education plan, brochures, fact sheets, presentation packets, slideshow and scripts, media kits, classroom presentation packets, and classroom learning stations. In addition, the Minnesota Extension Service developed an interactive HHM information station for public places.[59] According to Martindale[60] this "computer in a box" uses touch screens to provide users with a hands on approach to learning about HHM. Of the four units constructed, three were made available to counties, on loan. The fourth unit was placed in the state science museum for public education and outreach.

Example 2. Education Effort: Personal Contact

One-day HHM collection events in Iowa are termed Toxic CleanUp Days. The IDNR resolved that setting appointments for attendees of a Toxic CleanUp Day were an effective way to educate people. Using this approach, everyone that considers attending a Toxic CleanUp Day has to speak with a HHM specialist. It is easier and more efficient to learn about HHM one-on-one than en mass or in large groups.[55] For individuals calling in, an increased emphasis is placed on education, promoting the consumption of all material in a container, recycling and proper disposal of HHM.[60a]

Example 3. Consumer Emphasis: Incentive Through Financial Mechanism

As specified in Minnesota Section 325E.1151, Subdivision 1, a person who purchases a lead acid battery at retail must return a lead acid battery to the retailer or pay the retailer a $5 surcharge. This type of system is reinforced by the fact that placing a lead acid battery in mixed municipal solid waste is a crime, punishable as a misdemeanor.

Post-Consumer Phase

Post-consumer phase initiatives include traditional collection programs as well as financial mechanisms to generate operating funds for HHM programs. Several of these programs are described here.

Example 1. Collection Focus: Household Hazardous Material Collection Events

Amnesty Days was established as a three-and-one half-year project, implemented in six phases across all 67 counties in Florida.[61] The first of its kind in the country, this state sponsored mobile HHM collection program, allowed a participant, including homeowners, small businesses, government agencies,[62] and schools, to dispose of one barrel or 450 pounds of hazardous waste at one time. Brown[38] described the mobile collection program as tractor-trailers containing hazardous waste laboratories and storage space that traveled to designated sites within a county and stayed there for one to six days. Materials brought to the

site were sorted, packed, manifested, and eventually moved to a licensed hazardous waste storage facility. Items such as automotive batteries, motor oil, and paint
were often recycled locally.

According to the Amnesty Days Final Report, over 11,000 participants
brought in over 1.5 million pounds of hazardous waste in the first sweep of
Florida's counties.[61] The total cost was $2.8 million. The completion of the second
sweep of Florida's counties brought an additional 30,000 households, small
business, farmers, schools, and government agencies to contribute over 3.4 million
pounds of hazardous waste. Clarke[35] described the objectives of Amnesty Days
as threefold. First, Amnesty Days was to purge Florida of small quantities of
hazardous waste in storage. Second, it was to educate citizens on what hazardous
waste is and how that waste could be properly managed. Third, Amnesty Days
was a means to gain public acceptance for future local collection programs.

Example 2. Disposal Focus: Landfill Bans

In California lead-acid batteries, used oil, auto products such as antifreeze
and transmission fluid, household batteries, latex paint and all other HHM, as
well as mercury containing products, were banned from landfills.[65] Enforcement
of all bans is difficult.

Example 3. Disposal Focus: Financial Mechanism
of Solid Waste Taxes

Statewide, Minnesotans pay a 6% tax on garbage disposal.[66] This money is
directed to a state general fund, of which, approximately $1 million per year is
allotted to the MPCA to help run HHM programs.[60] Monies for administrative
and operating stipends to counties are at a flat rate. As more counties develop
permanent HHM programs, less money is available per county since the overall
amount of money available is constant.[43]

IMPLICATIONS AND FINAL THOUGHTS

Although the definition, terminology, and estimates of HHM in the municipal
solid waste stream vary, HHM concerns are real and indicative of increasing state
concern.

Implications of State Initiatives

The definition and amount of HHM that is generated is variable. Typically,
HHM represent less than 1% of the solid waste stream. Despite the differences
in estimates, taken in total, HHM represent millions of tons of material used and
disposed of each year. There is reason to believe that estimates for percentages
of HHM in the waste stream are low. Traditional methods for calculating percentages, such as waste composition studies, are limited in scope, and they do

not include materials that are stored in the home or discarded through other disposal routes such as in the backyard or down the drain. Also, as cities and counties recycle more non-hazardous materials, the percentage of hazardous materials may rise. Additional HHM information may be projected from manufacturing and product sales data.

Specific environmental contamination and human accidents relating to HHM are evident but not well documented. Exposures to hazardous materials are recorded each year by Poison Control Centers, and anecdotal information regarding accidents with municipal solid waste abound. Landfill issues encompassing municipal liability are real and can be financially damaging to a small or economically depressed community. The leaching of landfill liquids and chemicals has been established. But, the relation to landfilled HHM is not clearly substantiated due to the small quantities of HHM and the complexity in separating effects of small business wastes and other municipal materials. In contrast, incinerator ash and wastewater treatment concerns are well documented.

With respect to life cycle issues, the key for controlling HHM is threefold. First, pollution prevention and alteration of the production processes of manufacturers of selected HHM is crucial. Manufacturers should realize the impact of toxic product reductions and reformulations and act accordingly. Reducing container size and establishing consistent unit prices for various sizes of a similar product would allow consumers to buy only what they needed for a particular situation. This would minimize the need for disposal of unused product. Manufacturer reformulation specifications may force pollution prevention and foster cost saving strategies. Caution must be used in reformulation and development of new products as they may have deleterious effects as well. Realizing that manufacturers must make a profit, incentives to comply with regulations or standards are helpful. With respect to retailers, although retail permit fees are good in concept, they are an administrative nightmare. This type of system can only be accomplished through strong monitoring and compliance efforts. Retailers may be quick to pay $25 per year to appease enforcers but overall, only a limited amount of money is generated through this mechanism. Shelf labeling requirements for retailers are also difficult to monitor and enforce. An accepted definition of HHM would be needed, along with a number of trained people in the retail industry to identify and label products correctly. Another pre-consumer approach to limiting the amount of hazardous material is to charge the manufacturer or importer for the materials used and acquired. This is a method of fairly distributing the environmental burden of the hazardous material. If a manufacturer, wholesaler, retailer, or consumer chooses not to make, market, or use a particular material or product, then they are not impacted by the tax.

The second key to controlling HHM involves the consumer. A consumer must understand hazardous constituents are contained in the products they buy, store, and use, and understand the potential impact on the environment when materials are improperly disposed. Education and incentives that alter consumption habits are essential. Programs that provide consumers with a single person to discuss technical information allow direct communication between the homeowner and a HHM specialist. Financial incentives for recycling have worked well with items such as lead acid batteries.

The third way to control HHM deals with post-consumer collection and disposal. These efforts should minimize the impact to human health and the environment. Special HHM collection programs are expensive and reach only a small percentage of the population. They do, however, serve as an important educational tool. Landfill and incinerator bans are only as good as the incentives that affect compliance and enforcement, and although solid waste taxes generate large sums of money that can be channeled into HHM education and collection efforts, the tax itself is indirect.

Final Thoughts

State endeavors relating to the management of HHM are diverse in nature but are indicative of relative problems. Real and perceived risks (by citizens) have placed HHM management high on the agenda for many states. As a result, states are striving to become more innovative with the ways in which they deal with these material. The bottom line is to protect the health of people as well as the environment, while implementing cost effective and integrated programs and policies. A comprehensive and efficient state policy will target all phases of the life cycle and involve all stakeholders in the process towards solution and abatement of the HHM crisis.

REFERENCES

1. Hammond, A., Mast, T., and Silver, C., Eds., *The 1994 Information Please Environmental Almanac,* World Resources Institute, Houghton Mifflin Company, New York, 1993.
2. Litovitz T., Martin T., and Schmitz, 1986 Annual report of the American association of poison control centers national data collection system, *American Journal of Emergency Medicine,* 5(5):405, 1987.
3. Litovitz T., Schmitz, B., Matyunas N., and Martin T., 1987 Annual report of the American association of poison control centers national data collection system, *American Journal of Emergency Medicine,* 6(5):479, 1988.
4. Litovitz T., Schmitz, B., and Holm, K., 1988 Annual report of the American association of poison control centers national data collection system, *American Journal of Emergency Medicine,* 7(5):495, 1989.
5. Litovitz T., Schmitz, B., and Bailey, K., 1989 Annual report of the American association of poison control centers national data collection system, *American Journal of Emergency Medicine,* 8(5):394, 1990.
6. Litovitz T., Bailey, K., Schmitz, B., Holm, K., and Klein-Schwartz, W., 1990 Annual report of the American association of poison control centers national data collection system, *American Journal of Emergency Medicine,* 9(5):461, 1991.
7. Litovitz T., Holm, K., Bailey, K., and Schmitz, B., 1991 Annual report of the American association of poison control centers national data collection system, *American Journal of Emergency Medicine,* 9(5):452, 1992.

8. Litovitz T., Holm, K, Clancey C., Schmitz, B., Clark, L., and Odera, G., 1992 Annual report of the American association of poison control centers national data collection system, *American Journal of Emergency Medicine*, 10(5):494, 1993.

9. Environmental Protection Agency, Solving the hazardous waste problem: EPA's RCRA program, EPA/530-SW-86-037, Office of Solid Waste, Government Printing Office, Washington, D.C., 1986b.

10. Mooney, C., The Resource Conservation and Recovery Act and household hazardous waste, in *Proceedings of the Seventh U.S. Environmental Protection Agency Conference on Household Hazardous Waste Management*. Waste Watch Center, Boston, 1992.

11. Environmental Protection Agency, Project summary: management of household and small-quantity-generator hazardous waste in the U.S., Government Printing Office, Washington, D.C., 1990a.

11a. Environmental Protection Agency, Characterization of municipal solid waste in the U.S.: 1990 update — executive summary, EPA/530-SW-90-042A, Office of Solid Waste and Emergency Response, Government Printing Office, Washington, D.C., 1990b.

12. Mitchell, G. and DeMichelis, D., Handling household hazardous wastes, *Public Works*, (8):50, 1987.

13. Kleman, J., Florida Department of Environmental Protection, personal communication, 1993.

14. Iowa Department of Natural Resources, Household hazardous materials, in *DNR Code Extracts*, Des Moines, 1992, 455.

15. Rathje, W. and Murphy, C., *Rubbish! The Archaeology of Garbage*, Harper Collins, New York, 1992.

16. Fredrickson, L., Latham, C., Mitchell, S., and Thomas, J., Minnesota Pollution Control Agency Solid Waste Composition Study 1990-1991 Part I, Ground Water and Solid Waste Division, St. Paul, 1992.

17. Gorke, R., Hidden toxics in our homes: the household hazardous waste crisis, *Heal the Bay*, Santa Monica, 1991.

18. Kinman, R. and Nutini, D., Household hazardous waste in the sanitary landfill, *Chemical Times and Trends*, (7):23, 1988.

19. Duxbury, D., A look at HHW management trends, *Waste Age*, 20(10):80, 1989.

20. Ridgely, S., Hazardous waste from Minnesota households, Minnesota Pollution Control Agency, St. Paul, 1987.

21. Haugh, M., McCallum, L., and Green, W., Moderate risk waste — a progress report, volume 2-1 of the problem waste study, Washington State Department of Ecology, Olympia, 1990.

22. Galvin, D., Why household hazardous waste management is needed, in *Proceedings of the Fifth U.S. Environmental Protection Agency Conference on Household Hazardous Waste Management*, Waste Watch Center, Boston, 1990.

23. Darr, J., Cinalli, C., and Johnston, P., Screening consumer products for indoor air risks, *Proceedings of the Seventh U.S. Environmental Protection Agency Conference on Household Hazardous Waste Management*, Waste Watch Center, Boston, 1992, 76.

24. Mitchell, G. and DeMichelis, D., Handling hazardous household wastes, *Public Works*, (8):50, 1987.

25. Galvin, D., Why household hazardous waste management is needed, in *Proceedings of the Seventh U.S. Environmental Protection Agency Conference on Household Hazardous Waste Management*, Waste Watch Center, Boston, 1992, 63.

27. Sabel and Clark, Volatile organic compounds as indicators of municipal solid waste leachate contamination, *Waste Management and Research*, (2):119, 1984.

28. Ferry, S., The toxic time bomb: municipal liability for the cleanup of hazardous waste, *The George Washington Law Review*, 57(2):197, 1988.

29. *Facing America's trash: what next for municipal solid waste?* Government Printing Office, Washington, D.C., 1989.

30. Monz, D., Household hazardous waste management: minimizing costs and liabilities though regionalization, presented at the Eighth U.S. Environmental Protection Agency Conference on Household Hazardous Waste Management, Burlington, VT, November, 1993.

31. Bureau of National Affairs, Appeals court uphold Superfund liability for municipality in win for generators, *BNA Law Update*, Washington, D.C., Apr. 15, 1992.

32. Environmental Protection Agency, Characterization of products containing lead and cadmium in municipal solid waste in the U.S., 1970 to 2000 — executive summary, EPA/530-SW-90-042A, Office of Solid Waste, Washington, DC., 1989.

33. Environmental Protection Agency, Characterization of products containing mercury in municipal solid waste in the U.S., 1970 to 2000 — Executive Summary, EPA/530-S-92-013, Office of Solid Waste and Emergency Response, Washington, D.C., 1992.

34. General Accounting Office, Water pollution — nonindustrial wastewater pollution can be better managed, GAO/RCED/-92-40, Government Printing Office, Washington, D.C., 1991.

35. Clarke, R., A review of the Florida HHW collection program, in *Proceedings of the Sixth U.S. Environmental Protection Agency Conference on Household Hazardous*, Waste Watch Center, Ed., Boston, 1990.

36. Wexler, R., Interim project report of the water quality assurance trust fund — hazardous waste site cleanup: expenditure and revenue analysis, Florida Senate Committee on Finance, Taxation and Claims, Tallahassee, 1993.

37. Hall, B. and Kerr, M., *1991-1992 Green Index*, Island Press, Washington, D.C., 1991.

38. Brown, T., Household hazardous waste: the unresolved water quality dilemma, *Journal of Water Pollution Control Federation*, 59(3):120, 1987.

39. Florida Department of Environmental Regulation, Florida's groundwater program, *Florida's Environmental News*, 7(11):1, 1985.

40. Florida Department of Environmental Regulation, *Water Quality Assurance Act*, 83-310, 1983.

41. Iowa Department of Natural Resources, Toxic cleanup days — a report to the Iowa general assembly, Environmental Protection Division, Des Moines, 1987.

42. Krogulski, M., Iowa Department of Natural Resources, personal communication, 1993.

43. Minnesota Pollution Control Agency, Development of the permanent household hazardous waste management program in Minnesota, Hazardous Waste Division, St. Paul, 1991.

44. Goldsmith, L., Minnesota Pollution Control Agency, personal communication, 1993.

45. Gilkeson, J., Minnesota Pollution Control Agency, personal communication, 1994.

46. California, California Public Resource Code, Division 30, Part 7, Chapter 1, 1990.

47. Hughes, K., California Integrated Waste Management Board, personal communication, 1994.

48. Berton, F., California Integrated Waste Management Board, personal communication, 1994.

49. Green, W., Washington Department of Ecology, personal communication, 1993.

50. Winka, M., New Jersey Department of Environemntal Protection, personal communication, 1994.

51. Arnold, K., *Household battery recycling and disposal study*, Minnesota Pollution Control Agency, St. Paul, MN, 1991.

52. California Code of Regulations, Title 17, Chapter 8.5, Article 2, Consumer Products, 1992b.

53. California Code of Regulations, Title 17, Chapter 8.5, Article 1, Antiperspirants and Deodorants, 1992a.

54. Krogulski, M., Iowa Department of Natural Resources, personal communication, 1989.

55. Krogulski, M. and Gathright-Conner, C., Toxic cleanup days: a program that works for Iowa, *Iowa Conservationist*, (2):16, 1993.

56. Washington Department of Ecology, Model Toxics Control Act: 1993 Annual Report, WDOE 93-94, Olympia, 1993.

57. Drumright, M., Washington Department of Ecology, personal communication, 1994.

58. Goldsmith, L., Minnesota Pollution Control Agency, personal communication, 1994.

59. Gelbman, E., Regional reports: Region V, *Household Hazardous Waste Management News*, 2(7):5, 1990b.

60. Gelbman, E., Regional reports: Region V, *Household Hazardous Waste Management News*, 2(8):5, 1991.

60a. Martindale, M., Minnesota Pollution Control Agency, personal communication, 1994.

61. Laidlaw Environmental Services Inc., Household hazardous waste collection program issues information packet, Columbia, SC, 1992.

62. Environmental Protection Agency, A survey of household hazardous waste and related collection programs, EPA/530-SW-038, Office of Solid Waste and Emergency Response, Government Printing Office, Washington, D.C., 1986a.

65. Raymond Communications, *State Recycling Laws Update: Year End Edition 1994*, Raymond Communications, Riverdale, MD, 1994.

66. Brooks, N., Regional reports: Region V, *Household Hazardous Waste Management News*, 1(3):5, 1989.

67. Office of Technology Assessment, *Facing America's Trash: What Next For Municipal Solid Waste?*, Government Printing Office, Washington D.C., 1989.

CHAPTER 5

Composting: Programs, Process, and Product

Lynnann Hitchens and Richard M. Kashmanian

Composting is a microbial process in which organic materials are aerobically decomposed under controlled conditions to produce a humus-like product, compost. The composting feedstock can have a variety of sources: residences, restaurants and other commercial establishments, and agricultural sources, among others. The use of aerobic composting has become an effective landfill diversion tool for organic materials and a viable recovery and management option for municipalities. Composting systems of various types have become an important part of many integrated management systems.

Integrated resource recovery and waste management is the utilization of a variety of management options to achieve the goals of a particular community or geographic area. In designing an integrated system, EPA, the National Recycling Coalition, the Composting Council, and others, recommend the use of a hierarchy. This hierarchy (first published in the 1989 document The Solid Waste Dilemma: An Agenda for Action)[1] recommends a descending order of preference for management options. The most preferred option when feasible and advantageous is source reduction, or eliminating use of materials or toxic constituents in products. The next rung on the hierarchy is recycling, including centralized composting; third is combustion, preferable with energy recovery; and lastly, land disposal, necessary in all systems for materials that cannot be managed in other ways. An integrated system would not consist of only one management option, but a combination of options that make use of the hierarchy in a cost-effective manner.[1]

Composting is an integral part of the hierarchy, and is considered unique in that composting is an inherent part of both source reduction and recycling efforts. Backyard and on-site composting and the mulching of grass clippings are forms of source reduction. Composting of organic materials is considered recycling when utilized in a centralized manner. Regardless of how it is implemented, composting provides the production of a valuable product and increased landfill diversion. Increases in landfill diversion can be limited if only commodity recycling is relied on. Large increases in landfill diversion can only succeed with composting as a part of recovery efforts.

The residentially generated material beneficial to the composting process are the organic fractions: consisting mainly of yard trimmings, food scraps, woody materials, non-recycled paper, and some textiles. Other materials are essentially noncompostable, and are generally viewed as undesirable in the final product. There are other industrial and agricultural feedstock items sometimes used in composting programs, including food processing by-products, biosolids from municipal treatment operations, and agricultural and animal by-products.

There are many benefits to composting besides minimizing the amount of material that requires final disposal in a landfill and increasing diversion amounts, which can be important to public officials trying to meet state or local diversion or recycling goals. For example, compost has a variety of beneficial properties: compost provides a soil amendment that returns organic-rich material to the land, helps soil retain moisture and nutrients, increases soil fertility, reduces erosion, and soil compaction.

This chapter describes the organic feedstocks amenable to composting, the programs and systems currently in use, legislative laws and initiatives relating to composting, fundamentals of the composting process, and information about the compost product's utilization and marketing.

COMPOSTING FEEDSTOCK

The two municipal categories of material most amenable to composting are yard trimmings and food scraps. According to EPA, 32.8 million tons of yard trimmings and 13.8 million tons of food scraps were generated in 1993, accounting for 22.5% of the total amount generated from residential, commercial, and institutional sources.[2] Yard trimmings include grass, leaves, and tree and brush trimmings from residential, institutional, and commercial sources. Based on a limited collection of data, the EPA estimates that nationally, on average, yard trimmings are composed of 25% brush, 25% leaves, and 50% grass (by weight). However, it is recognized that these numbers vary widely depending upon the region of the country.[2] Food scraps include uneaten food and preparation scraps from residences and commercial and institutional sources.

Figure 5.1 depicts the division between compostable and non-compostable municipal discards. A maximum of 64% of the municipal discard stream is compostable. It is estimated that 6.5 million tons, or 3.1% of the municipal discard stream, was recovered for composting in 1993 (does not include backyard composting).[2]

Table 5.1 shows the composition of municipal discards by organic and inorganic classifications and the project growth in each of these areas. The organic

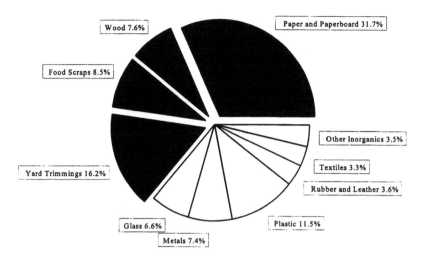

Figure 5.1 Compostable material as a percentage of municipal discards. Compostable materials are shaded. Note: A portion of the textiles may be compostable. Total quantity discarded: 161.95 million tons. (From U.S. EPA, *Characterization of MSW in the United States: A 1994 Update,* EPA/530-SW-94-042, November 1994b.)

component includes paper and paperboard, yard trimmings, wood scraps, and food scraps. The inorganic fraction includes all other components.

Certain grades of paper are suitable for both recycling and composting. There is often confusion existing as to what is the environmentally and economically preferable option. Recycling of paper into paper is generally considered the higher use option, though this solution may or may not be suitable for all municipalities. There are other circumstances, such as availability of recycling technology, markets, economics, and the ability to create a homogenous, contaminant-free feedstock, that may make one solution preferable over the others. Common planning guidance chooses recycling as the preferable option, where it is available, with

Table 5.1 Composition of Municipal Discards by Organic and Inorganic Fractions[a]

	Organic discards[b]		Inorganic discards[c]		Total
	Thousands of tons	Percent of total discards, %	Thousands of tons	Percent of total discards, %	Thousands of tons
1993	103,850	64	58,100	36	161,950
2000[d]	91,215	60	61,115	40	152,330

[a] Discards after recycling and compost recovery.
[b] Organic fraction includes paper and paperboard, yard trimmings, wood, and food scraps.
[c] Inorganic fraction includes glass, metal, plastics, rubber and leather, textiles, and other materials. Some textiles are recognized as compostable, but an exact amount is not available.
[d] Recoveries for 2000 are estimated at 30%.

From U.S. EPA, *Characterization of MSW in the United States: A 1994 Update*, EPA/530-R-94-003, May 1994a.

composting being used for soiled and non-recyclable paper. There is currently no estimate of the amount of soiled and non-recyclable paper available nationally.

EPA also makes national projections on residential generation, predicting an increase of 0.7% annually from 1990 through 2000. Materials such as plastic and paper will be increasing faster than average, and glass and food scraps will show slower increases than the average.[2] Estimated food scrap generation will be relatively constant, increasing from 13.8 million tons, or 6.7% of the total stream in 1993, to 14 million tons, or 6.4% of the total stream in 2000.[2] Yard trimmings generation is predicted to decline through the year 2000 due to significant source reduction expected to occur through mulching and backyard composting, education programs, legislation, and landfill bans.

It is estimated that 6.5 million tons of municipal discards (3.1% of generation) were recovered for composting in 1993. The outlook for composting as a management option is good; it is projected to increase to 11.2 million tons in the year 2000, or 5.1% of the generated municipal stream.[2] This amount does not include the material that is backyard composted. This illustrates the vast potential for composting that currently exists, and will still exist in the year 2000. The organic fraction includes biologically derived carbon-based materials that could be composted. It does not include chemically derived carbon-based material such as plastics. It should be pointed out that these total numbers are municipal discards, and projected recovery for recycling has already been deducted from these totals.

COMPOSTING PROGRAMS

Similar to the need for integrated management through source reduction, recycling, composting, combustion, and landfilling, an integrated strategy is needed for composting. That is, a mix of on-site (including backyard) and centralized composting and mulching systems should be considered. In addition, agricultural land application of uncomposted leaves and grass clippings should also be considered. To most effectively develop an integrated composting strategy, emphasis is needed on building the appropriate infrastructure.

One half of the states support backyard composting programs. These programs include education, promotion, funding, and/or compost bin distribution. In addition, there is support for backyard composting programs by local governments in over 80% of the states. Backyard composting programs exist in a range of community sizes.[3] Eleven states were able to estimate the number of these programs, for a total of 636.[4]

In nearly one half of the states, uncomposted leaves and/or grass clippings are applied to agricultural or horticultural land as a less costly method to manage these materials as compared to centralized composting and disposal. At least six states have guidelines for land application, including providing guidance on maximum time allowed for stockpiling, separation and removal of non-yard trimmings materials, depth and/or quantity of application, and timing of incor-

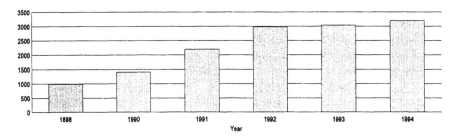

Figure 5.2 Yard trimmings compost facilities in the U.S. (Adapted from *BioCycle Annual Survey.*)

poration into the soil.[5] In addition, nearly two thirds of the states use mulching by "shredding or piling material with little management or biological activity" as one of their ways to recycle yard trimmings; in fact, "some of the 3,202 [compost] facilities are actually doing more mulching than composting."[4]

As illustrated in Figure 5.2, the number of yard trimmings compost facilities increased by roughly 50% each year between 1988 and 1992. However, as shown in Table 5.2, the number of facilities has stabilized since 1992, with 2,981 reported in 1992, 3,014 in 1993, and 3,202 in 1994.[4] In about two thirds of these states, these facilities compost grass clippings along with the leaves and/or brush.

Most of the compost facilities that handle municipal organics are operated by the local municipalities. However, a key role for the private sector (including farms) continues to emerge in operating compost facilities (on public or private land), providing oversight or troubleshooting services, marketing or brokering the compost, etc.[6] These services can offer benefits to municipalities by providing cost savings and greater composting experience and expertise. Privately owned and operated compost facilities are also more likely to handle municipal organics from multiple communities, providing potential economies of scale over smaller, individual sites.

Table 5.2 Number of Yard Trimmings Compost Facilities

Year	Number of programs
1988	651
1989	986
1990	1,407
1991	2,201
1992	2,981
1993	3,014
1994	3,202

Adapted from *BioCycle Annual Survey.*

There are at least 58 compost facilities (including on-site) where food service organics from grocery stores, hotels, quick service and full-service restaurants, and institutional cafeterias are composted, mixed with yard trimmings and/or other organic feedstocks.[7] In addition, farms are increasingly involved in composting municipal organics. For example, at least 24 of 46 farms listed in a directory for composting in Pennsylvania in 1993 mixed in yard trimmings.[7a] When compost facilities are allowed on farms, communities benefit because they do not have to incur capital costs for siting and constructing their own municipal compost facility, or a facility as large.

There are 18 centralized facilities that compost a mixed municipal feedstock, with 4 of these sites composting a source separated municipal organics stream (i.e., yard trimmings, food scraps, and non-recyclable paper). The other 14 facilities compost municipal organics and whatever recyclables and trash cannot be separated out at the facility from its incoming feedstock. The number of these facilities has not changed much during the past few years, with the number of new facilities somewhat balanced by the number closing.[8,8a]

LEGISLATION

States have taken various approaches to encourage and manage the development and oversight of composting programs. Examples of these approaches include:

a. incentives and assistance given for backyard composting programs (discussed above);
b. volume or weight-based pricing for curbside collection of discards;
c. guidelines for application of uncomposted yard trimmings to agricultural and horticultural land;[5]
d. composting manuals for municipalities to assist them in starting up or expanding their composting and mulching programs;[5]
e. compost standards and compost procurement guidelines and preferences to encourage the use of compost in public land maintenance activities on public lands;[5,9,10]
f. exemptions from permits for certain types of compost operations, e.g., backyard and on-site;[11]
g. a simplified process for registering or listing small leaf or yard trimmings compost facilities and farm compost operations (farms receiving off-farm materials may need to meet more stringent requirements);[11-13]
h. regulation of siting and management of larger leaf and grass, and source separated municipal organics compost facilities;[11-13] and
i. stricter regulations of siting and management for mixed municipal compost facilities.[12,13]

These measures tend to vary between states. For example, the regulations can include any of a number of requirements, such as where to site the facility in relation to surface and ground waters, wetlands, the flood plain, and residences and other human receptors; whether a permit is required; capacity of the facility; grade or slope of the land; orientation of the windrows to the slope of the land; size of the windrows; minimum windrow turning frequency; controls for leachate, dust, vectors, odors, noise, and *Aspergillus fumigatus*; access control; and visual screening.[11-13]

Currently, there are 23 states and the District of Columbia that have some form of landfill disposal ban on leaves, but in most cases the bans also include other yard trimmings such as grass clippings and brush.[5,14] Most of these bans are in effect — the remainder will be in effect by 1997. With 78% of the yard trimming compost facilities located in these 23 states, the implication is that disposal bans are effective in establishing compost facilities and diverting yard trimmings from disposal in landfills. Furthermore, if a state has a disposal ban, it is more likely to have also implemented multiple measures from the above approaches to develop an infrastructure to help manage its composting efforts.

A key lesson learned from these bans is that if they go into effect during the summer, only grass clippings would be available in large quantities for composting. As a result, malodorous conditions are very likely to result. These odor problems can result in closing the compost facility and/or rescinding or altering the ban. A better season to implement a disposal ban is in the fall, when leaves and brush are available for composting. In fact, if practical, when communities first get involved in composting, it is advisable that they take an incremental approach rather than compost all types of municipal organics at the start. First, communities could start composting leaves and then mulching brush, tree trimmings, and holiday trees and greens. Then over time, additional organic materials such as grass clippings, and perhaps later food scraps and non-recycled paper could be included in the recovery strategy. This incremental approach can avoid problems as:

1. the composting [and mulching] feedstocks become larger and more complex to manage;
2. the composting [and mulching] infrastructure expands;
3. the experience with composting [and mulching] grows; and
4. the confidence and acceptance of potential compost [and mulch] users increase for the compost [and mulch] product(s).[15]

Another lesson learned from these bans is that they need to be combined with education, assistance, incentives, and availability and access to alternatives.[15a] The establishment of new bans on landfill disposal of yard trimmings has tapered off substantially in the past couple of years. In addition, some bans have been scaled back, reducing or replacing restrictions on what yard trimmings are covered. In some states with no landfill disposal bans, municipalities or counties may impose their own bans. For example, in Texas, there are 18 local bans on disposal or collection for disposal of yard trimmings.[15b]

TECHNOLOGY

Technologies for composting vary from simple passive systems to the more sophisticated engineered systems. Regardless of the technology used, control of the composting process is dependent on the control of a variety of parameters such as oxygen, moisture, and carbon-to-nitrogen ratio. In the presence of oxygen and moisture, organic materials will decompose. From this basic premise, composting systems are designed to deliver oxygen, moisture, and nutrients to accelerate and optimize this microbiological process. While the individual systems in existence are too numerous to mention, categorically, there are three general methods for composting: turned windrow, static pile, and in-vessel.

The turned windrow method utilizes a windrow, or pyramid-shaped pile, typically 5 to 6 feet high and 10 to 12 feet long. The typical rule of thumb is a windrow's length should be twice the height as the basic composting structure.[16] The windrow's shape allows for ease of turning, or mixing, and adequate exposure to oxygen and moisture.

The windrow can be turned in a number of ways, from using a front-end loader, to a specially designed "windrow turner" that passes through the windrows, with tines agitating, aerating, and reducing the particle size of the material. Without turning and sufficient aeration, the composting process is slowed and can result in substandard compost quality. The turning of the windrow is necessary to ensure that all of the material spends an adequate portion of time in the interior of the pile, where the faster decomposition takes place. The turning of the windrow becomes the single most necessary physical parameter, as the turning frequency and efficiency have an impact on important compost parameters such as oxygen, temperature, and moisture content.

The static pile method uses the windrow or pile structure, but piping systems are placed underneath or within the windrow. Oxygen is delivered to the decomposing material by forcing air through the material, or applying suction to pull the surrounding air through the material. Aerated static piles can be 10 to 12 feet high,[16] minimizing the space needed for composting. In the static pile method, physical turning of material is less of a necessity. The advantage of applying a suction system is the minimization of odors released to the atmosphere. Air pulled through the static pile can be directly vented to an odor control system, or biofilter.

The static pile method is most suited for material of a relatively uniform particle size, and a relatively dry feedstock. This will minimize "short-circuiting" of the air flow which would allow anaerobic pockets to develop in areas of excessively wet material, or material of different sizes that would have a tendency to clump. Bulking agents such as sawdust or wood chips can be added to minimize these effects.

The in-vessel type of system can take on a number of different configurations depending on design and vendor technology. The majority of in-vessel systems consist of a relatively short composting period in the vessel which accelerates the decomposition process, coupled with a significantly longer curing phase. Inherent to the in-vessel system is the mechanical control of the major composting parameters, such as oxygen, moisture delivery, and feedstock mixing. In-vessel systems depend in part on the type of materials composted. In-vessel systems generally require a greater capital investment, and compost program designers must determine what the balance is between composting time, variable controls, and available capital and operating monies.

PROCESS

Successful composting can be defined as the creation of a clean, marketable, and stable compost. The foundation of successful composting is the control of several fundamental parameters: oxygen, moisture, and carbon to nitrogen (C:N) ratio and the monitoring of others, such as pH, and temperature.

Oxygen

For successful microbial decomposition, there must be adequate oxygen to support the dominance of the aerobic microorganisms. Without adequate oxygen, anaerobic conditions can occur, leading to odor problems and a poor compost product. The aerobic microorganisms are continually consuming oxygen, and as the supply decreases, decomposition can slow. The agitation, turning, or mixing of the compostable material is necessary to increase pore space within the windrow or pile. In addition to dissipating heat and water vapor, aeration recharges the oxygen supply, increasing the rate of decomposition. Aeration can be passive or forced, depending on the composting technology used.

Moisture

The presence of water is necessary to the composting process, however the balance of water-to-solids is a delicate parameter that can have a large impact on aerobic activity. For the purposes of composting and organic feedstock analysis, moisture content is expressed as the weight of water as a percentage of the total or wet weight of the material.

Moisture content = [(wet weight − dry weight)/wet weight] × 100

The literature reports a variety of ranges for optimum moisture content. This number greatly depends on the type of feedstock, its particle size, and the rate and type of composting desired. Generally, ranges of 50–60% are desirable, recognizing that moisture contents above 65–70% interfere with desired oxygen levels.[16] If the moisture content drops below 45–50% it will interfere with microbial activity.[17]

C:N Ratio

There is a significant amount of data and information in the literature recognizing the importance of the carbon-to-nitrogen ratio (represented as C:N) to the

Table 5.3 Carbon-to-Nitrogen Ratio of Various
          Materials

Type of feedstock	Ratio
High carbon content	
Bark	100–130:1
Leaves and weeds (dry)	90:1
Mixed MSW	50–60:1
Paper	170:1
Wood	700:1
High nitrogen content	
Cow manure	18:1
Food scraps	15:1
Fruit scraps	35:1
Grass clippings	12–20:1
Leaves (fresh)	30–40:1
Mixed grasses	19:1
Biosolids	11:1
Weeds (fresh)	25:1

From EPA, 1994.[16]

composting process. Attainment of an adequate ratio will produce significant biological activity and minimize the potential for odors. A ratio between 25:1 and 35:1 is generally agreed upon as optimal for composting.[16] In the case of a heterogeneous feedstock, the combination of many materials of varying C:N ratios can be balanced to produce a mix with an overall ratio in the desirable range. If there is too much carbon, biological activity will decrease; if there is too much nitrogen, the excess will be released as ammonia, both a source of odors and toxic to microorganisms.

Grass clippings are a good source of nitrogen with a C:N ratio in the range of 12–20:1; leaves generally have a C:N ratio of 90:1.[16] Yard trimming compost programs in geographic areas with four distinct seasons are well familiar with the unique set of problems that comes with a seasonal feedstock. During the spring and summer months, grass is the predominant yard trimming and it alone has a C:N ratio lower than the optimal range. This necessitates an adjustment with the addition of a material with a higher C:N ratio. The natural complement to grass is leaves and woody materials (e.g., brush), the majority of which are generated in the fall. Some composting programs have found that conserving the leaves over the winter until the months in which grass clippings are generated produces a more balanced C:N ratio. The generally recommended ration of grass to leaves in approximately 1:3 to balance the carbon and nitrogen content. Table 5.3 presents the C:N ratio for a variety of organic feedstocks.

pH

During composting, the pH will follow a progressive pattern, corresponding to the type of microbial activity taking place. The pH will drop during the initial stages of the composting cycle as microorganisms break down the carbonaceous materials and produce organic acids. The synthesis of organic acids is accompanied by the development of a population of microorganisms capable of utilizing the acids as a substrate.[17] This will cause the pH to rise. Microorganisms have an optimal pH range of 6 to 7.5. Fungi have a wider optimal range, from pH 5.5 to 8[18] and can more easily tolerate changes in pH. It should be noted that if the pH rises to 9, nitrogen is converted to ammonia and becomes unavailable to microorganisms.[19]

Temperature

Like pH, temperature is not usually a controlled variable, but is an indicator of the microbial activity existing in the decomposing mass. Both mesophilic and thermophilic organisms are necessary for successful composting and these organisms are naturally present in organic material. Mesophilic microorganisms grow best at temperatures between 25 and 45°C.[20] As the microorganisms metabolize the organic matter, generating carbon dioxide and water, the temperature of the composting mass rises. Under less than optimal temperatures, the mesophiles become dormant. Thermophilic microorganisms prefer temperatures between 45 and 70°C.[20] So, as the temperature of the composting pile rises, the thermophilic microorganisms dominate. The phase in which the thermophiles are generating heat is the point at which pathogens are destroyed. Time and temperature are necessary to ensure that pathogens are destroyed. Thermophilic decomposition continues as long as sufficient nutrients and oxygen exist.

After the active composting phase, it is necessary to have a curing phase. The purpose of a curing phase is to ensure that the compost is stabilized, allowing the remaining available nutrients to be metabolized by the available microorganisms. During the curing phase, decomposition of the material continues, but oxygen is consumed and heat is generated at much lower levels. If an immature compost is utilized, it can continue to consume oxygen and reduce the availability of oxygen to plant roots.[19] Immature compost can also contain high levels of organic acids that can be damaging for certain applications. It is recommended that the curing process should continue for a minimum of one month.[19]

COMPOST PRODUCT

Compost has a range of benefits: it releases nutrients slowly, helps the soil to retain moisture and nutrients, improves soil fertility and tilth, reduces soil erosion and soil compaction, and suppresses plant disease, etc.[9,21] There are various uses for compost that take advantage of these benefits, such as soil enhancer; organic-based fertilizer; top dressing and topsoil mixes; residential gardens; potting soil mix; plant disease suppressant; erosion control; stormwater runoff control; biofilter; livestock bedding; soil renovation, reclamation, and remediation; and landfill cover.[9,22] For most of these uses, a consistent quality and quantity of product are necessary to penetrate and retain a market segment or user.[23]

Mulch, which in some cases can be considered a compost that is not fully mature, also has a range of benefits, including moderating soil temperature, suppressing weeds, and reducing soil erosion, soil spattering, and the spread of soil-borne disease.[21] "Shredding woody materials can produce various grades (e.g., fine and coarse) of mulch. The composting process can be used to prepare woody and vegetative materials into a better mulch product, and in less time, than it would take to produce a humus product (i.e., a mature compost). With the high temperatures achieved in composting, weed seeds and plant diseases can be inactivated or killed. In addition, the decomposition will darken the color of the mulch produced, and more closely resemble commercially available grades of mulch."[9]

There are several markets for compost: agriculture, landscapers, nurseries, public agencies, and residents;[9,22] similarly for the use of mulch by landscapers, nurseries, public agencies, and residents. These markets seek varying levels of product quality.[9] However, the larger markets, such as crop agriculture, tend to pay a lower price for compost.[24,25] Currently, agriculture lags behind the other markets, except in cases where farms produce the product themselves. When the yard trimming compost facility is operated by the municipality, the compost is primarily used by public agencies and residents.[25a]

Ideally, potential users of the compost (or mulch) to be produced (and their product needs) should be identified before the compost (or mulch) facility is first opened. Due to the high cost of transportation relative to the value of compost (and mulch), a 25 to 50 mile radius around the compost (or mulch) facility should be targeted for marketing and distributing the product.[26] "However, the actual distance would depend on the quality and value of the compost, form of sale (i.e., bag or bulk), access to transport arteries, and type and size of transport vehicle."[9] As an example of the types of tradeoffs in answering these questions, bags of compost command a higher unit price than the compost sold in bulk, but bagging compost carries with it more equipment and/or operational costs.

This product/market information from potential users will determine the type (e.g., compost or mulch) and quality of the product sought by the facility. The product/market information will also influence whether compost or mulch will be produced; the feedstock(s) to be processed; the quality and quantity needed for the compost (or mulch); pre-processing, processing, and post-processing needs; product distribution needs (e.g., bulk or bag distribution, pick-up or delivery); whether new equipment needs to be purchased or leased; development of a product name; pricing structure and its relationship to product distribution; types of educational and product promotional materials to distribute; and whether research or "field" demonstrations are needed to advance the flow of information about the benefits of using compost (or mulch) and how to use it. Much of this information can come from focus groups prior to facility construction or other types of feedback from the green industry.

Some communities have offered compost free or at reduced prices initially to attract users and markets. However, others in the composting industry feel this tends to devalue the product in the customers' mind (making it difficult to later charge a price) and recommend that it be sold at a positive price to cover at least some of the processing and/or transportation costs. Whether or not a price is charged may also depend on whether the composting facility is a public or private operation.[9]

The successful marketing of compost product depends not only on the physical quality of the product, but also the chemical quality. Excessive levels of trace metals, toxic organic chemicals, and inorganic substances are unacceptable to both regulators and compost users. At the national level, only compost containing biosolids as a part of the feedstock has regulatory requirements. Other composts, such as those made with yard trimmings and food scraps, do not have national regulatory requirements. There are however, a number of states (e.g., Florida, Maine, Massachusetts, Minnesota, New Hampshire, New York, North Carolina, and Vermont) that regulate all or most types of compost. The most commonly regulated metals include cadmium, chromium, copper, lead, mercury, nickel, and zinc.

There are a number of preprocessing and screening steps that can be taken to minimize metals in the final product. However it has been found that the lowest levels of heavy metals were consistently achieved by source separation of compostable material.[27] This would eliminate potential feedstock items that would contribute to the levels of heavy metals and other undesirable substances in the product. As with many other environmental issues, source reduction is a desirable option.

One type of composting expected to become more important is the composting of source separated organic materials derived from residences, and commercial and institutional sources. Through a growing number of pilot programs and a number of full-scale programs, it has been discovered that the prevention or exclusion of certain problematic materials up front has improved compost product quality, and ease of operation.

There are numerous data on heavy metals in compost, and several literature evaluations of metals in final product.[10,16,27,28] The values presented are often highly variable, both between studies and within the same study; because of this, regional and operational trends are not obvious. A number of cautions should be taken when evaluating literature data:

- The trace metal and organic contaminant values presented are usually total concentrations, which can only partially define risk. There are other parts of the equation — bioavailability and mobility of contaminants.
- Sample digestion and analysis methods are not always discussed in the presentation of data, and differences can affect the final totals.
- Often single samples are taken and used inappropriately, without adequate attention to quality assurance and quality control.
- Foreign matter (glass and plastic, etc.), stability, and maturity are not often measured.

Since predictive measures are not appropriate, site specific sampling and analysis plans are necessary on a case-by-case basis.

SUMMARY

The composting industry is reaching a crossroads as well designed and operated programs become sustainable programs, and others fail for a variety of reasons. Concerns about odor, product quality, and marketing will continue to be important to vendors, operators, and the general public. These concerns will spur technological development through research beneficial to the composting industry as a whole. This will ensure the success of composting as a viable technology.

REFERENCES

1. U.S. EPA, *The Solid Waste Dilemma: An Agenda for Action,* Office of Solid Waste and Emergency Response, EPA/530-SW-89-019, February 1989.
2. U.S. EPA, *Characterization of MSW in the United States: A 1994 Update,* EPA/530-SW-94-042, November 1994.
3. Johnson, H., Backyard Composting Education Programs, *BioCycle,* 36 (1), January 1995, pp. 75–77.
4. Steuteville, R., The State of Garbage in America: Part I, *BioCycle,* 36 (4), April 1995a, pp. 54–55, 58–63.
5. Kashmanian, R., Composting and Agriculture Converge, *BioCycle,* 33 (9), September 1992, pp. 38–40.
6. Glenn, J., Private Managment of Yard Waste, *BioCycle,* 30 (1), January 1989a, pp. 26–28.
7. Kunzler, C. and Roe, R., Food Service Composting Projects on the Rise, *BioCycle,* 36 (4), April 1995, pp. 64–71.
7a. Oshins, C. and Graves, B., *On-Farm Composter Directory,* Rodale Institute Research Center and The Pennsylvania State University, Draft, October 1993.
8. Goldstein, N. and Steuteville, R., Solid Waste Composting Plants in a Steady State: Part II, *BioCycle,* 36 (2), February 1995, pp. 48–50, 52–56.
8a. Goldstein, N. and Steuteville, R., Solid Waste Composting Seeks Its Niche: Part I, *BioCycle,* 35 (11), November 1994, pp. 30–35.
9. U.S. EPA, *Markets for Compost,* Office of Solid Waste and Emergency Response, and Office of Policy, Planning and Evaluation, prepared in conjunction with CalRecovery, Inc. and Franklin Associates, Ltd., EPA/530-SW-90-073A, November 1993.
10. Bye, J., Setting Standards for Compost Utilization, *BioCycle,* 32 (5), May 1991, pp. 66–70.
11. Glenn, J., Regulating Yard Waste Composting, *BioCycle,* 30 (12), December 1989b, pp. 38–41.
12. Goldstein, N., Guidelines and Permiting for Yard Trimmings Composting: Part I, *BioCycle,* 35 (7), July 1994a, pp. 62–65.
13. Goldstein, N., Strategies for Regulating Source Separated Organics Composting: Part II, *BioCycle,* 35 (8), August 1994b, pp. 55–57.

14. Kashmanian, R., Quantifying The Amount of Yard Trimmings To Be Composted In The United States In 1996, *Compost Science & Utilization,* 1 (3), Summer 1993, pp. 22–29.
15. Kashmanian, R. and Spencer, R., Cost Considerations of Municipal Solid Waste Compost: Production Versus Market Price, *Compost Science & Utilization* 1 (1), 1993, pp. 20–37.
15a. Steuteville, R., Measuring the Impact of Disposal Bans, *BioCycle,* 35 (9), September 1994, pp. 58–60.
15b. Steuteville, R., The State of Garbage in America: Part II, *BioCycle,* 36 (5), May 1995b, pp. 30, 32–37.
16. U.S. EPA, *Composting Yard Trimmings and Municipal Solid Waste,* EPA/530-R-94-003, May 1994.
17. *The BioCycle Guide to Yard Waste Composting,* The JG Press, Inc. Emmaus, Pennsylvania, 1991.
18. *The BioCycle Guide to The Art and Science of Composting,* The JG Press, Inc. Emmaus, Pennsylvania, 1989.
19. Rynk, R., et al., *On-Farm Composting Handbook,* Cooperative Extension, Northeast Regional Agricultural Engineering Service, Ithaca, NY, 1992.
20. Atlas, R. M., *Microbiology,* 2nd ed. MacMillan Publishing Co., New York, 1988, pp. 438–440.
21. Kashmanian, R. and Keyser, J., The Flip Side of Compost: What's In It, Where to Use It and Why, *The Environmental Gardener,* Brooklyn Botanic Garden, Handbook #130, 48 (1), Spring 1992, pp. 15–20.
22. Alexander, R. and Tyler, R., Using Compost Successfully, *Lawn & Landscape Maintenance,* November 1992, pp. 23–24.
23. Alexander, R., Expanding Compost Markets, *BioCycle,* 31, 8, August 1990, pp. 54–59.
24. Tyler, R., Fine-Tuning Compost Markets, *BioCycle,* 35 (8), August 1994, pp. 41–42, 47–48.
25. Tyler, R., Diversification at the Compost Factory, *BioCycle,* 34 (8), August 1993, pp. 50–51.
25a. Cox, K., Expanding Markets for Yard Waste Compost, *BioCycle,* 30 (10), October 1989, pp. 64–65.
26. Cal Recovery Systems, Inc., *Portland Area Compost Products Market Study,* Prepared for Metropolitan Service District, October 1988.
27. Richard, T. L. and Woodbury, P. B., What Materials Should be Composted?, *BioCycle,* 35 (9), September 1994.
28. Hyatt, G. W. and Richard, T. L., Eds., *Biomass and Bioenergy.* Pergamon Press, Tarrytown, New York, 3, 1992, pp. 3–4.

CHAPTER 6

Scrap Tire-Derived Fuel: Markets and Issues

Michael H. Blumenthal

INTRODUCTION

This chapter reviews the current use of scrap tire-derived fuel (TDF); describes some of the reasons why TDF has become an attractive fuel; discusses processing methods of scrap tires; and identifies the issues often cited when the use of whole or processed TDF is discussed.

Background information on the Scrap Tire Management Council (STMC) and its approach to the scrap tire situation is also presented. The STMC is a non-profit organization sponsored by the North American tire manufacturers. The Council is an advocacy organization, whose main focus is to identify and promote environmentally and economically sound markets for scrap tires. The Council supports all markets for scrap tires which meet these two criteria. This presentation, however, focuses on only one of the three basic markets for scrap tires.

OVERVIEW

Each year, this country generates approximately 242 million scrap tires, or approximately one scrap tire per capita. Since the Council has been in existence (October 1990), the number of scrap tires having markets has increased from 11% (25 million) to approximately 37% (90 million) of the annual generation rate. The Council estimates that there are approximately one billion scrap tires in stockpiles around the country, although there is no data to support this estimate. The Council is in the process of collecting information on stockpiles from the states, but not all states have this information.

1-56670-215-1/97/$0.00+$.50
© 1997 by CRC Press, Inc.

TDF MARKETS

Of all the scrap tire markets, the TDF market is the largest single market for scrap tires, consuming 80 of the 90 million scrap tires that have markets. This market can be broken into seven segments:

1. cement kilns;
2. pulp and paper mill boilers;
3. utility boilers;
4. industrial boilers;
5. foundry cupolas;
6. resource recovery facilities; and,
7. dedicated scrap tire-to-energy facilities.

The total TDF market segment has the capacity to consume some 200 million scrap tires a year. The tire-derived fuel (TDF) requirements range from whole tires, to rough shreds, to a $1'' \times 1''$ shred, to a $2'' \times 2''$ shred. The type of TDF necessary for any end user will be a function of the type of combustion facility, the type of fuel used, and the feeding system used. For the purposes of this chapter, the term "TDF" will be used to describe all fuel applications for scrap tires. When appropriate, a description of the fuel used will be provided.

SCRAP TIRE PROCESSING

In certain cases, whole scrap tires can be used as fuel. This use is perhaps the best fuel application, since it obviously requires no processing and because the end user typically will receive a tip fee for accepting the tire. When processed TDF is required, the scrap tire must be size reduced to meet the end user's specification. The processing of a scrap tire is not a one-process event. Depending upon the size particle desired, a scrap tire can go through several different phases and types of processing equipment before it is ready for its market application.

While this presentation describes a variety of equipment and a system for processing scrap tires, it should be kept in mind that there is no one specific way to process scrap tires. The information presented herein is intended to give a general overview of processing equipment, and does not suggest or advocate a preferred processing system. The various phases of scrap tire processing can be described as follows.

First phase processing reduces the whole scrap tire to a "one pass" or "rough shred" material. The size of the first phase shred varies greatly depending upon a series of factors including, but not limited to, the condition of the knives and the rate at which the tire passes through the knives. Typically, however, one can expect a shred 3–6 inches (75–150 mm) in width by 3–12 inches (75 mm to 300 mm) in length.

The most frequently used machines used in primary shredding are slow-speed shear shredders. These shredders have been specially designed for size reducing scrap tires, and there are many manufacturers offering a variety of products. Shear shredders usually consist of an in-feed hopper and feed holding area. The in-feed area is typically located on top of, or feeds into the shredding area, which consists of two, counter-rotating shafts which often are designed to rotate at different speeds. Hammermills can also be used, although they are not common in this phase of the processing. This type of processing is also referred to as an "ambient grind" or "ambient processing," since the processing of the tire takes place at the temperature of its immediate surroundings.

In some tire processing operations, the bead wire is removed before the scrap tire is shredded. This device, aptly referred to as a "de-beader," removes the two sections of the tire containing the bead wire. (The primary function of the bead wire is to assist in forming a seal along the tire's rim. This traps the air in the tire, which eliminates the need for inner tubes.) While this added step certainly will make shredding easier, it adds another expense and additional time to the process.

The uses for the shred produced in the primary processing phase are very limited. In general, the only use that this type of shred can be applied towards is as a low-grade fill material. Even under the best of circumstances, handling of this material is difficult due to the quantity of exposed bead and belt wire, which causes the shreds to interlock and "nest" (form large, roughly shaped wads). The market for primary shreds as TDF is limited, due to the handling difficulties.

In the secondary size reduction phase, the rough shred is sent to another machine via a belt, where it is size reduced to a smaller chip (e.g., 2–3 inches in width and length, or 52 mm–75 mm). Once again, slow speed, shear shredders are commonly used in this phase of processing, although other ambient processing equipment can also be used. Shredders have the advantage that they can handle a fairly high volume of material, but have a limited belt and bead wire removal efficiency.

The other equipment used during this phase of processing includes hammermills, granulators, or crackermills. All three of these technologies have a better wire removal efficiency than shredders because they can generate smaller-sized particles, which enables the magnets to remove a greater percentage of wire. While more efficient in wire removal, there are other trade-offs. For example, hammermills have a greater energy demand and maintenance costs than shredders. Granulators, which are also referred to as knife mills, can handle comparable capacities as a shredder but are not specifically designed for scrap tires. Crackermills are common in scrap tire processing, but have a limited volume capacity.

Once the size reduction process is completed, the TDF can be delivered to an end user. One question that can be asked is: what are the advantages of using TDF?

BENEFITS OF TDF

TDF offers two advantages to the end users. The first advantage is cost, since processed TDF is less than the price of coal on either a per ton or per million

BTU basis. If a facility can make use of whole tires, then that facility will receive a modest tip fee for accepting those tires. Tip fees at the various types of facilities that use whole tires range from $0.15 to $0.50 per tire. Clearly, there are cost savings that can be realized from the use of TDF. This feature makes TDF combustion very attractive to many plant managers.

The second competitive edge which TDF offers concerns emissions. In facilities obtaining less than 10% of their fuel from TDF have normally not experienced any significant differences in their emissions. Those facilities which have 10% or more of their fuel derived from scrap tires, at worst, have shown no increases, and at best, have actually realized a net reduction in emissions.

There are three reasons most often cited for the decrease in emissions when TDF is used. First, TDF has 25% greater heating value than coal, so their tends to be a more complete combustion of the fuels. Second, tires are a very consistent material. Chemically, there aren't many differences between the various brands. Furthermore, tires contain fewer elements than most coals. This assists in the net decrease in emissions because there is more heat being released while fewer elements are being combusted. Third, in facilities that have an electrostatic precipitator (ESP), the metals emitted from scrap tires tend to coat the exchange sites in the ESP, increases the charge which, in turn, makes the system more efficient.

COMBUSTION SYSTEMS

In order to get scrap tires into any of these facilities, it will be necessary to determine whether the facility has the a combustion system capable of accepting TDF, whether TDF is compatible with the fuel being used, and whether the proper pollution control equipment is in place.

From the aspect of combustion systems, TDF can be used in virtually any cement kiln, cyclone, wet bottom, or grate boiler system. For example, TDF is used in all the various cement kiln technologies in this country (long wet, long dry, pre-heater, pre-calciner, and pre-heater/pre-calciner). In pulp and paper mill boilers, TDF is used in several stoker fired grate systems. Coal-fired utility or industrial boilers using TDF either have a stoker grate, cyclone or wet bottom boiler. The dedicated scrap tire-to-energy facilities have a specially designed grate system.

The types of TDF used also varies according to the combustion technology. For example, coal fired boilers with stoker grates, cyclones or fluidized beds for utilities, industrial or pulp and paper mill boilers all use processed TDF. Coal-fired wet bottom boilers can use whole tires. In the cement industry, whole or processed TDF, in theory, can be used interchangeably in virtually any kiln. Field applications indicate there is a preference for whole tires, although some kilns actually fire both processed and whole tires at the same time.

Table 6.1   Nationwide TDF Use

	Cement kilns	Pulp and paper mills	Electric generating[a]
Using TDF	26	10	11
Trials conducted	11	10	11
Considering TDF	25	2	3

[a] All inclusive.

Costs to convert facilities to tire-derived fuel are typically not the limiting factor. The length and cost of the permitting process, as well as the availability of a consistent, quality supply of TDF and the capacity of the air pollution control technologies will usually determine whether a facility can or will convert to the use of TDF. At present, there are some 62 facilities across the country that are currently in the permitting process or seriously considering the use of TDF. Perhaps the easiest manner to determine whether the facility in your area is interested in the use of TDF is to contact them. Information about current TDF usage (see Table 6.1) can be obtained from the U.S. EPA or the Scrap Tire Management Council.

Another benefit associated with the use of TDF is that it provides a market for the billion scrap tires (estimated) stockpiled around the country. The reason for this is straight forward — stockpiled scrap tires accumulate water, dirt, sand, rock, and the like. While this dirt has little to no impact on the end use/end users, this "dirty" scrap tire poses many problems for the scrap tire processor. Putting a "dirty" scrap tire through a processing system will cause increased wear on the cutting knives. Dulled knives (or other cutting surfaces) produce a low quality TDF, while replacement of the knives is very costly.

LIMITING FACTORS

There are limits, however, to many of the TDF markets referred to. In cement kilns, TDF use is unlikely to exceed 25% of the total fuel stream. This is due primarily to the characteristics of the materials being used to make cement, although zinc is the main limiting factor. If cement contains more than 4,000 parts-per-million, its setting time will be delayed, which is undesirable.

In pulp and paper mill boilers, the limit on TDF appears to be 10% of their fuel mix. The reasons generally cited are the effects any wire remaining in the tire chip may have on the grate. There have been some cases where the slag (melted bead wire) has been welded to the grate, making its removal difficult.

In utility boilers, the limits to TDF are generally a function of the impact the slag might have on the coal ash. Utility boilers, in general can sell their bottom slag as an aggregate. However, TDF use in utility boilers is limited to two to five percent of the fuel mix. While this is a relatively small percentage, in actual terms this may consume a significant number of scrap tires.

ISSUES ASSOCIATED WITH TDF

While the TDF market is increasing, it is not without its opponents. Perhaps TDF's greatest problem is one of perception — the belief that the combustion of tires simply must yield terrible black smoke, noxious odors, and toxic emissions. To a certain extent, one can understand the rationale for this — uncontrolled, outdoor tire fires are known for these characteristics, and it is simply assumed that if you have environmental problems with one, there will be environmental problems with any form of tire combustion. Reports by the U.S. EPA confirm the Council's position that the use of TDF has actually improved emissions.

CONCLUSION

There are several significant markets for scrap tires. At present, the largest single market for scrap tires is fuel, either whole or processed. The type of tire-derived fuel consumed is a function of the combustion process and feeding system. The likely potential market for TDF is 200 million scrap tires a year, which represents a potential share of 80% of all the scrap tires generated annually.

CHAPTER 7

The Use of Scrap Tires in Rotary Cement Kilns

Michael H. Blumenthal and Edward C. Weatherhead

INTRODUCTION

The Scrap Tire Management Council was established in 1990 by the North American tire manufacturers. The Council's primary objective is to assist in the development and promotion of markets for scrap tires that meet two criteria. First, the use (or market) must be environmentally sound, and second, economically viable. In 1990 the Council sponsored an evaluation of markets for the use of scrap tires.

Two markets, both having large-scale consumption potential were identified. These markets were as a supplemental fuel (whole or processed tires) and as ground rubber for asphalt pavement. The largest potential fuel market identified in this evaluation was cement kilns. To expand the use of scrap tires in cement kilns, the Council has prepared this report that describes the use, destruction, and energy recovery of whole or processed scrap tires in rotary kilns (for the purposes of this chapter, whole or processed scrap tires used as a supplemental fuel will be referred to as TDF — tire-derived fuel).

This presentation may not answer all technical aspects relative to the use of TDF. It does, however, supply information and suggest strategies that can answer many of the initial questions concerning the use of TDF. Following are suggested procedures based on field experience and conversations with kiln operators using TDF, and are not representative of any given technology. The actual procedure or protocols used at a facility may vary, due to environmental requirements or different conditions at a plant.

If, after reading this chapter, there are additional questions, the Council would be glad to arrange a visit in which the technical, regulatory, and supply aspects of this matter can be further discussed.

1-56670-215-1/97/$0.00+$.50

MANUFACTURE OF CEMENT

The raw materials for cement consist of limestone (calcium carbonate), clay (alumina), sand (silica), and iron (in almost any form). These materials are blended in an exact recipe in accordance with the chemistry of each component. Cement is made by injecting the raw materials into a cement kiln and heating the materials to a temperature range of 2,650 to 2,750 degrees Fahrenheit (°F). At this temperature the formation of tricalcium silicate, or alite, the principal compound of Portland cement clinker, occurs. To arrive at this temperature, a flame temperature of 3500°F is necessary.

All kilns are inclined, higher at the raw meal feed end than at the discharge end. Kilns incline from the feed end to the discharge end at one quarter inch to one half inch per foot of length. Kilns revolve at a rate of one to three revolutions per minute. The combination of the kiln incline and the revolving cause the raw feed and clinker to move through the kiln. (The schematic drawing in Figure 7.1 illustrates a preheater precalciner kiln.)

A time/temperature situation is necessary for the clinkering phenomena to occur. A positive oxygen atmosphere is also necessary for quality clinker to form. Fuels are injected into the low end of the kiln (Figure 7.1, No. 4) and the gases of combustion flow upward to the discharge end. Thus, the gases of combustion and the raw meal flow in opposite directions. The clinker falls out of the kiln at the lower, or discharge, end of the kiln into a clinker cooler. The clinker cooler, while cooling the clinker, conveys it to a transport system that takes the clinker to a storage area. From the clinker storage, the clinker is transported along with about 5% gypsum, by weight, to a "finish" grind mill. Other additives included at this point facilitate grinding or make a special type of cement. The cement next passes through an air classifier and either returns to the mill for further grinding, or goes to one of several storage silos.

Production Of Cement

Production rates may be increased in preheater kilns while burning whole tires. This is possible by virtue of the preheater calcination rate increasing in the preheater second, third, and fourth stages when burning tires, compared to the normal calcination rate while burning coal only. Calcining rates have been increased from 45% burning coal only to 56% when burning whole tires. Use of tires has decreased the carbon dioxide transported by the kiln which, in turn, allows room for additional oxygen to be used in the kiln. The extra oxygen allows for the burning of additional clinker.

All of the above kilns require clinker coolers to receive the burned cement clinker and cool it, and to preheat the primary, secondary, and tertiary air flows that are used to burn the clinker, dry the coal, and to help calcine the raw meal.

As the calcium carbonate heats in the preheater of the kiln, it calcines when it is in the 800 to 2200°F range (calcium carbonate becomes calcium oxide plus carbon dioxide). The calcium oxide is lime, and the carbon dioxide is the gas that flows out the stack. The chemicals formed, which make up cement clinker,

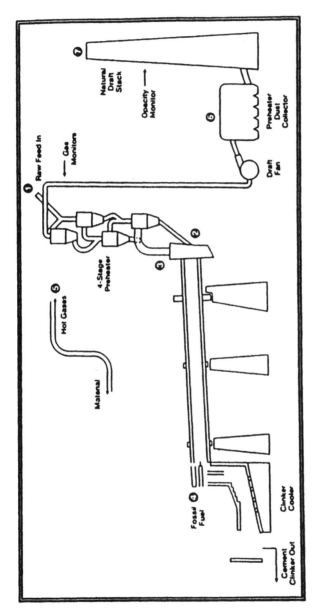

Figure 7.1 Typical preheater cement kiln.

are tricalcium silicate, tricalcium aluminate, and tetracalcium aluminoferrite. The gases of combustion and dust particulate are drawn through the kiln and through the air pollution control device (APCD) by induced draft fan(s). The APCD may be a cyclone (multi-clone), an electrostatic precipitator, a bag house, or a combination of any two.

KILN CONFIGURATION

There are five basic cement kiln configurations:

1. Long dry kilns
2. Long wet kilns
3. Four stage preheater kilns
4. Preheater-precalciner kilns
5. Grate preheater kilns (Lepol kiln)

Long Dry and Long Wet Kilns

These kilns have chain systems hung near the raw meal feed end that facilitates the transfer of heat from the kiln exhaust gases to the raw meal. The chains are hung in either garlands or in curtains so that they can readily absorb the heat from the exhaust gases. These kilns range in diameter from 9 to 24 feet, and in length from 350 to 725 feet. Long dry or long wet kilns may use either whole tires or shredded tires as a supplemental fuel. Shredded tires can be fed into the kiln by insufflation, that is, blowing shreds into the discharge end of the kiln. Due to the short residence time (three to five minutes) within the kiln, the shreds should be two inches or smaller, which assures the complete combustion of the TDF prior to falling into the clinker cooler. Another method of tire shred insertion into the kiln is by a chip cannon. This equipment can handle up to five tons of various sized chipped or shredded scrap tires per hour.

There are patented systems currently available in the U.S. that allow the feeding of whole tires into the kiln's calcining zone. These technologies could allow for an extended service life of this group of kilns, whose high fuel costs can make these facilities marginal producers of cement. (Further information on these patents and/or other related equipment companies can be obtained from the Council.)

Preheater, Preheater-Precalciner, and Grate Preheater Kilns

These kilns range between 9 and 17 feet in diameter, and 160 to 265 feet in length. Preheater-precalciner kilns may burn either shredded and/or whole tires. Two-inch chips may be fed with the coal in a precalciner. Whole or shredded tires can be fed as well in between the fourth stage and the kiln at the riser duct to the fourth stage preheater vessel.

Tires can be introduced at the riser duct to the fourth stage preheater vessel through a double tipping valve (Figure 7.1, No. 3). This tire feeding system, designed at the Southwestern Portland Cement (SPC) facility at Fairborne, OH is now in use by several cement kilns in the U.S.

The system consists of an elevator rising to the level of the top of the kiln at the feed end. The whole tires are discharged from the elevator to a gravity conveyor. This conveyor crosses a weigh scale that sends a 4 to 20 milliamp signal to the kiln control computer. The computer then slows or speeds the elevator to decrease or increase the feed rate of tires to the kiln. The whole tires travel down a short piece of gravity conveyor after passing the scale. This short piece of conveyor contains a photoelectric cell. When this cell's light path is broken, it triggers a double tipping valve. This is a series of two valves used to retain the negative air pressure inside the kiln. At the end of the short piece of conveyor noted above, the first of the two valves opens. The whole tire drops by gravity through the valve. This first valve then closes and a second valve opens, allowing the whole tire to drop down a chute onto the feed shelf. The momentum of the drop and the slope of the feed shelf cause the tire to actually bounce and roll into the kiln.

When this system type was first installed around the country, whole, scrap passenger tires were the most common tires used. Recently, however, many kilns have begun, and actually prefer, whole truck tires in this system. While the opening must be larger than for passenger tires, the increased weight of truck tires imbeds the larger tires more deeply into the meal. This allows for a more even distribution of heat throughout the meal, as well as a greater source of heat.

In addition to the Southwestern facility in Fairborne, this system has also been used by Southwestern Portland at Kosmos Cement in Louisville, KY; Medusa in Clinchfield, GA; Calaveras in Redding, CA; and, Lafarge in Whitehall, PA.

Destruction of TDF

Scrap tires (TDF) can be completely destroyed in cement kilns for a variety of sound technical reasons. The combination of extremely high temperatures (2650 to 2750°F), a positive oxygen atmosphere, and a relatively long gas residence time (4 to 12 seconds at the elevated temperatures) assures the complete combustion of the scrap tire. The complete combustion precludes products of incomplete combustion (PIC) or black smoke or odors being released from the stack.

TIRE-DERIVED FUEL

Characteristics of TDF

Eighty-eight percent of the tire is composed of carbon, hydrogen, and oxygen, which accounts for its rapid combustion and relatively high heating value. Tires contain approximately 15,000 BTUs per pound. This compares favorably to coal which, on the average, contain some 12,000 BTUs per pound. Subsequently, when substituting TDF for coal, a kiln operator can reduce coal by 1.25 pounds for every pound of TDF used.

An additional advantage of TDF use is its steel portion. A 20-pound passenger car tire contains two and one-half pounds of high grade steel. The steel can substitute, in part, for the iron requirement in the cement raw meal recipe.

Another point of interest is that tires tend to have a lower percentage of sulfur than most coals. Sulfur in tires ranges from 1.24 to 1.30% by weight. Sulfur in coal ranges from 1.1 to 2.3% or higher, depending on the coal quality. The average coal used in cement manufacturing will run approximately 1.5% sulfur. Calcium carbonate, the largest single ingredient in cement, is one of the most effective natural sulfur gas scrubbers. The presence of calcium carbonate helps control sulfur emissions from a cement kiln. Emissions data from a variety of kilns has clearly demonstrated a consistent reduction in sulfur and other emissions with the use of TDF (Figure 7.2a–2d). Since all the components of the tires are either destroyed, combined into the clinker or captured in the APCD, there is no ash disposal. Finally, the components of the scrap tire, once chemically combined

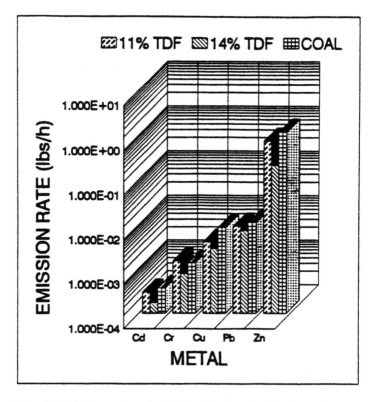

Figure 7.2a Analysis for metals using EPA method 12 during stack testing for Holnam Incorporated Industries. Arsenic was below detection limits.

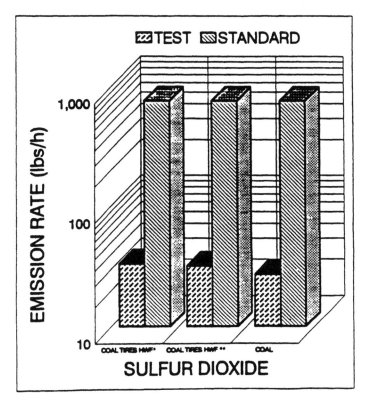

Figure 7.2b   Gaseous sulfur emissions from April 1991 Ohio compliance stack test at Southwestern Portland Cement Company. * = Higher metals input rate. ** = Lower metals input rate.

into the clinker, are not capable of leaching out. This is comparable to silica not leaching from glass. In short, the cement kilns use 100% of the scrap tire in an environmentally sound manner.

Cost Considerations

Another significant advantage to using whole tires is that it can lower operating costs, as compared to using 100% coal. Currently, whole scrap tires can be obtained by an operating facility at a positive cost. That is, a tip fee can be assessed by the kiln operator to scrap tire transporters in exchange for the right to deliver scrap tires to the cement company. The use of scrap tires reduces the tonnage of coal used, and consequently lowers the cost associated with the acquisition of coal. Finally, as indicated in the preceding section, the steel component of a tire can substitute for iron, thereby reducing the cost of iron acquisition.

The use of shredded scrap tires, while not as cost effective as whole tires, typically can be obtained for less than the cost of most coals. Since no new

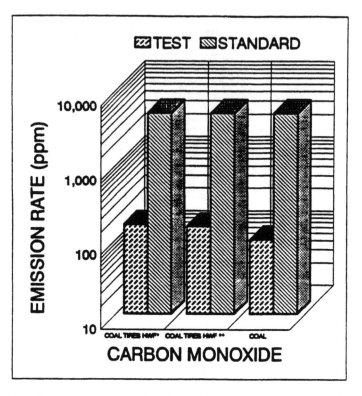

Figure 7.2c Gaseous carbon emissions from April, 1991 Ohio compliance stack test at Southwestern Portland Cement Company. * = Higher metals input rate. ** = Lower metals input rate.

pollution control devices are required, the only capital expenses required are a trailer storage area and a feeding system. As has been demonstrated by several of the kilns, the costs related to the use of TDF are similar to any improvement project in a cement plant. The payback period for the capital improvements is generally less than 18 months, depending upon the percentage and cost of the TDF used. The actual capital costs to construct a tire feeding system will vary, depending on the kiln configuration and technology used and the complexity of the system itself. Generally, these capital costs have ranged between $200,000 and $500,000.

Actual operating cost should change very little. The one area that may increase is the additional manning requirement. Field experience suggests that the sophistication of the feeding system affects the number of persons required to load and operate the tire feeding system. The very complex, and obviously more expensive, feeding system requires only one individual, if any at all, to operate it. As the complexity of the system is reduced, the manning requirements increase. Maximum personnel should be four to five persons total. In the case of a one person operating per shift, this equates to 4.2 persons when considering three-shift day,

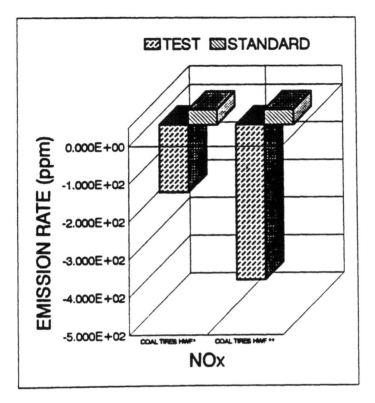

Figure 7.2d   Gaseous nitrogen emissions from April, 1991 Ohio compliance stack test at Southwestern Porland Cement Company. * = Higher metals input rate. ** = Lower metals input rate.

seven-day-per-week format. Quality control may need an extra 0.5 person (day shift only) to handle the extra testing suggested for the first six months of TDF use.

When adequate experience has been gained, the need for the extra quality control/quality assurance should diminish. Suggested parameters requiring observation during testing include raw materials, fuels, clinker dust, crystal structure, cement color, grindability, setting times, and cement strengths (including but not limited to masonry, which may have shorter board life due to the higher strength cements).

Maintenance costs are likely to increase slightly. There should be a lower coal mill maintenance due to the reduced coal throughput; however, there will be maintenance required on the TDF feeding/weighing system that may offset this reduction. Another initial expense encountered has been the cost to train plant personnel with the new feeding equipment as well as the expense of additional fire fighting equipment near the TDF conveying feeding system. The plant's fire fighting plan should also be modified to reflect the additional fire protection systems installed for the TDF system.

CEMENT QUALITY

The quality of the cement, whether using whole or shredded tires and regardless of the point of entry, is as good or better than when using only coal. Holnam's Ideal Cement facility in Seattle, WA, which has long straight kilns, has provided better quality cement using TDF than with their basic fuel of a coal and petroleum coke mixture. Holnam uses two-inch tire shreds at a feed rate of 2.5 tons per hour. The shreds are injected into the kiln at the discharge end by insufflation. They estimate the shreds are lofted 25 to 30 feet into the kiln, and never had evidence of tire steel in the cooler. The clinkering zone was shortened and alite size was made smaller, and the clinker crystal definition was improved.

The improvement feature is the stabilization of their free lime. Prior to tire burning, free lime jumped around considerably. This stopped with the burning of scrap tires due, the company believes, to a shorter clinkering zone brought on by the high volatility and high energy content of the TDF.

The cement quality has also improved in preheater kilns when using TDF. This is brought about by two basic facts:

1. The raw meal calcining in the preheater is increased by 22 to 25%, depending on the tire burn rate. This, in turn, has made the kilns operate more smoothly, which makes for more consistent quality.
2. Crystal structure and clinker appearance have both improved; clinker size is also reduced.

The combination of these improvements has been demonstrated at one plant in a two-day grindability test on Type I cement. The ground cement's Wagner increased to 2140, as compared to the preceding month's 2000 Wagner average. The Blaine was 3860 compared to 3640 average for the preceding month's average Blaine. The percent of clinker passing 325 mesh increased from 94.6% average over the preceding 30 days to 96.9% for the two-day test grind. From field experience, it is reported that there was no discernible cement color change. Furthermore, cement quality was as good or better than base line studies and the microscopy studies revealed more discreet crystal structure.

There has not been any reported buildup of rubber in the fourth stage preheater vessel. There have been reported additional buildups of calcined raw meal scale in the area of the feed shelf, which was easily managed. Buildup on the feed shelf, which is just up from the feed end of the kiln, has occurred in a small number of preheater or precalciner kilns. The normal buildup consists of lime scale or raw meal scale and is frequently removed by water blast guns or air cannons. The buildup of TDF on the feed shelf can occur due to the angle of the feed shelf being either too flat, or due to the tires not being fed in a manner that allows the inertia of the tires to carry them towards into the kiln.

Other Considerations

Historically, there has been a limit to the quantity of TDF used in a cement kiln. Heidelburg Cement (Germany), which was one of the first kilns to use TDF, has not exceeded 25% fuel substitution with TDF. There appears to be a sound reason for this limit — the combined zinc content of the TDF and other fuels may have an impact on cement quality.

TDF contains approximately 1.5% zinc. If the total zinc content of all the fuels exceeds 4,000 parts per million (PPM), there may be an increased setting time. The impact of a increased setting time is obvious. Consequently, an analysis of the total fuel stream would be beneficial to insure that zinc levels remain within an acceptable range.

Plant Operations

Some additional points to consider when first burning TDF follow. When starting up a tire burning system, operators frequently overfuel with TDF. While all reasons this occurs are unknown, the primary reason is that TDF contains 1.25 times the BTU value of the comparable weight of coal. Overfueling is detrimental to clinker quality, cools the kiln, and wastes fuel.

Close attention to CO gauges may spot the problem. Another certain way to recognize a reducing atmosphere is to analyze the belite crystalline structure. A yellowish ring around the edge of the prepared crystal is an indicator of overfueling the kiln. Another consideration is overfluxing, a situation caused by too much iron. The symptom for this would be a color change in the cement.

KILN PRODUCTION RATES

The production rates in long, straight kilns and preheater–precalciner kilns, at least according to the operators we have spoken with, were unchanged. Production rates in preheater kilns may well increase by virtue of calcining completed in the preheater from 20 to 25%. Increased production rates may be realized if there is fan capacity for additional oxygen, or there is cooler capacity for additional clinker, or if there is feed and grind capacity for additional raw meal. This really should not be surprising since it is the same principle that preheater-precalciner kilns use to increase production over preheater units of like size.

There have been concerns raised regarding the addition of tires at the feed end of the kiln and the possible effect of depriving the discharge end of the kiln of the heat necessary to have good clinkering action in the burning zone of the kiln. Preheater–precalciner kilns frequently burn as high as 65% of their fuel in the precalciner and 35% at the discharge end of the kiln. When clinkering occurs, there is an exothermic reaction. That is, the clinker puts out heat during formation. Thus, the fact that 65% of the total fuel may be injected 250 feet upstream from where the clinkering takes place has little or no effect on the clinkering action.

In the case of preheater kilns, where 25% of the fuel is taken from the discharge end of the kiln and injected into the feed end, is a "spit in the bucket" compared to what occurs in preheater–precalciner kilns, where 65% of the fuel from the discharge end of the kiln is taken. It should also be stated that when using TDF in precalciner kilns, fuel must be removed from the precalciner to offset the TDF injected into the feed end of the kiln.

TEST METHODS

Stack Testing

It is suggested that companies notify the appropriate governmental agencies of their intent to burn TDF; test the stack prior to burning TDF; and request a meeting to discuss the permitting and testing protocol. At a minimum, the agencies to contact are the air quality group and the solid waste management group having jurisdiction over your area. Each of these agencies should be given time to discuss this subject with their departments. Many questions may arise at this stage for which the answers may require research. The Scrap Tire Management Council is available to assist in this effort. Agencies can be reassured that similar facilities are using TDF, and that these companies, and the Council, can furnish a listing of permitted facilities.

Permit forms should be sought and permitting requirements should be reviewed for future use. (It is imperative that the agencies receive the information concerning the plant's intentions from the plant manager, not from a newspaper reporter or through rumor.) Permit requirements will vary with each state. Most permits will be in two distinct parts: air quality and solid waste. The air quality permit may be a revision of the current kiln operating permit. The revision may be required due to the addition of a new fuel, but the use of TDF should not be considered as a new source of emissions. The second part of the permit would comply with the state's solid waste material handling laws. It should be noted that scrap tires are not a hazardous material.

Air Quality (AQ) Permit Considerations

Suggest to the regulatory agency that the tests for air quality emission standards are based on practices used in other states, standards developed by the U.S. EPA, or on existing data. The following test protocol is suggested:

particulate, CO, SO_x, NO_x, metals of interest to the state regulatory agency, total unburned hydrocarbons (THC), and hydrogen chloride (HCl).

Destructive removal efficiency (DRE) is a test usually associated with the burning of hazardous waste fuels (HWF), but occasionally, regulators consider TDF in the same category as HWF when it comes to stack tests, so you may have to run a DRE. To do a DRE, a surrogate for hydrocarbons must be "spiked" or added to the fuel to prove the destructive removal efficiency usually to 99.99% or better removal.

A surrogate that is easily detectable in the analysis process, stable at high temperature, and not normally found in any of your raw materials or fuels should be selected. Freon™ 113, which is easy to detect but difficult to destroy, may be the best surrogate for this use (it vaporizes at 118°F and is easily identified on a gas chromatograph).

Many states require a HCl test which, until a test other than that presently prescribed for incinerators is promulgated, should be contested as the test is absolutely unfair. The test, as designed, measures total chloride and then assumes that it is all HCl. This may be true for incinerators or other industrial furnaces, but it is virtually impossible for lime or cement kilns to emit HCl gases due to the calcium carbonate atmosphere in the kiln. Calcium carbonate is present in quantities exceeding 10-fold of all fuels burned in the kiln. The calcium carbonate is finely ground and is a gas scrubber unequaled in any type of furnace. What may be caught by stack testers are salts of chlorine such as KCl, NaCl, CaCl, none of which affects litmus paper, much less forms HCl when dissolved in water.

The subject of lead vapor going out the stack arises far more frequently in the burning of hazardous waste in cement kilns than it does when burning TDF. However, when using TDF in a kiln that is also burning hazardous waste, especially a chlorine-laden hazardous waste, the problem of lead vapors going out the stack is very real.

The vapor pressure of lead is normally 1160° F at 1×10^{-6} atmospheres (Table 7.1). Should chlorine gas be present at a 10% level, the vapor pressure goes to 5°F at 1×10^{-6} ATM. A 10% chlorine presence is very high — much higher than normally encountered — but it is a straight-line function of the percentage of chlorine present as to the volatility of lead. Thus, the presence of chlorine in a kiln system is very important, especially if lead is also present.

Considerations for the Test Burn

Environmental authorities are usually very anxious to run stack tests on cement kilns soon after the kiln begins using TDF. The reasons for this expedited testing schedule have no scientific or technical basis. All things being equal, it is advisable to run TDF through the kiln system for several days before beginning any testing program. The kiln system usually requires this time to become stabilized, which yields more accurate results.

The amount of time required for a kiln system to become stable varies depending upon the process. In the case of burning TDF, the point that requires the longest time to settle down is the injection of the heat source near the rear or feed end of the kiln instead of at the discharge end of the kiln. This action changes the heat profile at the discharge end and the feed end of the kiln. Some of the variables affected include the change of calcine rates, change in brick coating, change in kiln rings, and change in the point where clinkering occurs. Each of these variables may take anywhere from three days to three weeks to stabilize. Furthermore, settling the kiln operators into a new set of operating parameters takes, at best, one to two weeks. Thus, 30 days should be the minimum amount of time a kiln is operating with TDF before any testing program begins.

After receiving a permit to conduct a test burn, a series of two or more short-duration test burns is normally scheduled. These tests should be designed to test conveying and feeding equipment, the effect on the kiln burning, clinker quality, and to test for any raw meal recipe changes which may be necessary. After the short-term tests have been concluded and appropriate adjustments made, a long-term test, 60 to 120 days, should be scheduled (the longer, the better). During this test burn, a stack test may be performed in which complete quality testing should be done. Kiln brick life should be closely observed during the long term test as well, and permitting agencies and local decision makers should be invited in to observe the project. In regard to product quality, microscope studies should be done at least daily to observe crystal size and definition, alite and belite formation; and crystal reactivity should be observed.

Other Considerations

Each time a whole scrap tire enters the kiln system at the feed end of the kiln, a CO spike occurs if the CO sample inductor is too close to the feed end of the kiln. This spike occurs due to the volatility of the rubber. This rubber grabs all available oxygen at the point of entry. There is additional oxygen present on up and down the kiln so the CO meter quickly stabilizes when the tire is being combusted. There is a time window that many environmental authorities allow wherein you may exceed the allowable CO level for a second or two, but in the preceding and following seconds, the CO must run a specific number of points below the maximum permissible levels to satisfy the permit conditions.

Suggested Sampling

The following is a suggested list of the kiln system points that should be sampled at half hour intervals throughout the entire stack test, and at normal plant procedure sampling intervals for the balance of the test period. Sample points should include, at a minimum, the kiln raw meal or slurry feed, and kiln dust from APCDs, both from the main and alkali bypass units if appropriate. Clinker, pulverized coal, and coke at the discharge end and at the precalciner, TDF, and hazardous waste fuel should be sampled, if being used. If some test results look suspect, more frequent sampling may be necessary.

During the stack testing, samples taken at the one-half hour intervals should be composited into four hour samples for testing. This is done to assure representative samples are used for testing. The various samples should be tested as follows: kiln feed, kiln dust, and clinker should be analyzed with the use of an X-ray fluorescence spectrophotometer for silica, aluminum, iron, calcium, magnesium, sulfur, sodium, and potassium. Microscopic examination should be conducted on all composited samples throughout the TDF test burn and especially during the stack test.

Samples should also be sent to a commercial testing laboratory to test for heavy metals and for toxicity characteristic leaching procedure (TCLP), as many state environmental agencies require these data. The test for heavy metals should include arsenic, barium, cadmium, chromium, lead, mercury, selenium, and silver. It is wise to add zinc since all tires contain this element. Fossil fuels should be tested for ultimate analysis and mineral ash content plus TCLP on the heavy metals. TDF should be tested for moisture, ash, volatiles, fixed carbon, BTU, sulfur, nitrogen, hydrogen, and oxygen. Each test should be run on an as received basis and a dry basis.

Furthermore, the following suggested list of parameters should be recorded during the TDF testing: raw meal or slurry feed rate; coal feed rate and percent of total fuel; coke feed rate and percent of total fuel; TDF feed rate and percent of total fuel, clinker production; and, dust production.

The kiln shell temperature profile should be recorded hourly for six hours before the start of the test and then every four hours throughout the test to check for brick damage, coating profile changes, etc. Kiln scanners are satisfactory but hand-held optical pyrometers at ten foot stations along the length of the kiln are better. Each set of readings should be graphed. All process monitoring and their recorders should be in top working order for the entire test period. Any facility not having recorders should rent recorders for the test period. Stack opacity monitors, when not required, are also suggested.

The importance in this rests on the fact that regulators frequently use CO as the ultimate measuring stick as to your ability to completely burn TDF. So, watch carefully for overfiring in the first days of each kiln operator's first experience using TDF, as a reducing atmosphere will yield high CO readings. Most regulators require a base line test prior to your TDF stack test. Try to have kiln operating conditions in a "normal" operating status during the base line tests.

The following point has only arisen in cases when HWF was being burned in conjunction with TDF; however, it could occur without HWF. The presence of chlorine in the kiln or stack can cause some metals to volatilize at far lower temperatures than normally is the case. If these temperatures are present beyond the dust collection equipment, the metals leave the stack and enter the environment. This, of course, can be cause for failing a stack test if the metals escape in quantities that exceed existing regulations. The volatility of certain metals is not affected by the presence of chlorine (i.e., chromium, beryllium, barium, antimony, selenium, cadmium, arsenic, and mercury). Table 7.1 shows volatility of some metals that are affected.

Table 7.1 Volatility of Selected Metals

Metal	Volatility* temp (°F)	Volatility* temp (°F) with 10% Cl
Nickel	2210	1280
Silver	1660	1160
Thallium	1330	280
Lead	1160	5

* The temperature at which vapor pressure is 1×10^{-6} atm.

SOLID WASTE PERMIT CONSIDERATIONS

Handling Scrap Tires

For most facilities, this aspect of the permitting process should be a minimal effort, especially when using whole tires. Whole tires should never hit the ground, as they should only arrive, be stored, and fed off an enclosed trailer, or from a live bottom hopper. Normal fire prevention/fighting measures can be modified to include the management of TDF. In the case of processed TDF, following the suggested procedures in the Tire Supply section should suffice for the solid waste management requirements. M.S.H.A. (Manufacturer's Safety Data Sheet) health standards should be closely observed. Truckers delivering tires must be trained in the plant's safety procedures involving their activities.

Tire Supply

When testing begins, or the announcement of intent to do a test burn, tire suppliers will "come out of the walls." All potential suppliers should fill out a questionnaire furnished by the cement company, asking questions such as:

- Will a manifest system be used to track scrap tires from their sources to the cement plant, complete with weights and signatures from suppliers, truckers, and cement plant personnel?
- Can the specified tires size(s) be furnished?
- Will clean, off-rim, water-free, vermin-free tires be furnished?
- Will tires be furnished in closed top vans?
- Will any and all tires be taken back that are not in compliance with the above criteria at no cost to the plant operator?
- What is the price FOB cement plant? (Prices on a per pound basis.)

Perhaps the most significant question that has to be answered by the prospective TDF suppliers is whether they have an adequate tire supply and the appropriate state or local licenses/permits. If a supplier does not have the required documents, he should simply not be considered, regardless of his tire supplies or offering price.

Once all of these questions have been addressed by the prospective TDF suppliers, the next phase of negotiations should begin. Field experience provides us with excellent qualifying contract conditions. We suggest that the prospective supplier be required to supply a performance bond to the plant manager. This bond can be in the form of a certificate of deposit, a letter of credit, or other mutually agreeable financial instrument. The two basic purposes of this requirement are to separate the pretenders from the real players, and to give your organization the assurances it needs to make such a commitment.

Whole and processed TDF both require special consideration when receiving or storing them on site. Whole tires, due to their shape, can retain up to two gallons of water and are ideal breeding grounds for mosquitoes and rodents.

In order to avoid these potential storage problems, it is suggested that when using whole tires only, covered trailer load shipments be accepted. Furthermore, there should be no reason to remove the whole tires from the on-site trailers except to off-load them onto a conveyor feed system leading to the kiln. This simple management decision may alleviate the need for a solid waste management permit since no scrap tires can be dumped or loosely stored on-site. Obviously, the trailer in which the whole tires are stored must be an intact, closed-top van trailer. The handling requirements for processed scrap tires is very similar to those of coal. Processed TDF cannot retain water, nor does it offer a haven for mosquitoes. The potential for fire is minimal, since tires will not auto-ignite until 1800°F. About the only significant management concern would be rodent infestation. However, timely pile turnover and the absence of all potential food sources in the general proximity of the pile should maintain the pile rodent-free.

A fire prevention/fire fighting plan should be devised in cooperation with the local fire department. The Council has developed guidelines for the management of scrap tire fires, *Guidelines for the Prevention and Management of Scrap Tire Fires,* which includes prevention, pre-incident planning, and fire fighting tactics. This document and a training course are available from the Council.

Specific routes for the trucks delivering the tires should be developed and a plan given to the drivers. The truck route should be developed such that it does not interfere with normal kiln operations. Since the use of TDF will decrease the quantity of coal used, this change may increase the heat at the coal mill. It is important to check the cooling capacity of this system to preclude any overheating at the coal mill.

INTERNAL CONSIDERATIONS

Once the management decision has been made to test TDF, it has been found helpful to inform the employees and the public of the proposed TDF project. This generally helps to keep the rumors and misconceptions down to a minimum. Thus, a public relations effort is necessary. To begin the program, plant employees and the public should be given a continual supply of information about the plant's intentions. Topics of interest will include, but are not limited to, what the goals are and how the company will accomplish these goals; what the permitting status is; which other plants are burning TDF (and how they go about it); where tires are coming from; how they are transported; the routes over which the transporters travel; and how this fuel will affect air emissions.

Communications

A company spokesperson should be appointed in addition to the plant manager so that in any event, someone will be available to answer questions relating to the use of TDF. We suggest the following communication items be considered:

Employee understanding — explain through memos or other standard forms of communication that the facility will soon begin testing TDF as a supplemental fuel. Points of interest should include the manner in which the TDF will be used, the fact that no black smoke will be produced and that the use of TDF will benefit both the plant (through lower fuel costs and increased productivity) and the community (by supplying a market for the scrap tires).

Customer Relations

Have the sales force or management explain the technical details of TDF use to your customers. Of particular interest should be the point that the use of TDF will have no adverse effect on the quality of the cement. You may want to explain that, in certain cases, cement strength has improved with the use of TDF. This same point, the fact that there are no adverse effects to product quality, should also be explained to the employees.

Employee Safety

One of most common misconceptions about the use of TDF is that it produces odors, black smoke, and toxic emissions. Employees may be concerned about the effects to their personal health due to the possible toxic emissions. The point that tire combustion in the kiln will not produce toxic emissions should be stressed. Data are available to clearly demonstrate that the use of TDF has actually lowered criteria emissions. It may be helpful to have the plant's union leadership contact union leaders at those plants using TDF for reassurances.

Public Relations

As with your employees, the public will likely have the same misconceptions (emissions, odors, and black smoke). The education of the community may be your greatest concern, but one that can be addressed effectively. There are several ways to address this issue, and hiring the services of a local, professional public relations firm is suggested. Notwithstanding this decision, there are nine items that should be considered.

1. Let the public know early on that you are considering the use of TDF.
2. Repeat this message often.
3. Hold an open house, let the public see what you intend to do.
4. Answer their questions and address their concerns.
5. Hand out printed material on this subject (a Q&A brochure is available from the Council).
6. Send representatives to every public forum possible to communicate the possible benefits (providing a market for scrap tires — more markets mean fewer and smaller scrap tire piles. Emissions are improved, no black smoke, and that the facility is working with the state and local environmental regulatory agencies). Cost savings, while important to the plant, are usually of little concern to either the public or the regulators.

7. Publish a monthly newsletter and distribute it widely.
8. Send news releases to the local newspaper.
9. Send four or five community leaders to a plant burning tires to see how the system actually works.

CONCLUSIONS

This chapter has reviewed some of the key and more common issues raised when a facility first considers the use of TDF. From field experience it can be concluded that tires can be used successfully in cement kilns, and that good quality cement can be made from the clinker generated while using scrap tires. Furthermore, production rates can be increased and fuel costs can be lowered while burning whole scrap tires. Environmental quality can be maintained or improved using scrap tires in cement kilns. Finally, use of scrap tires in cement kilns is a viable energy recovery program. The Scrap Tire Management Council offers technical assistance to kiln operators further explaining this material and related issues. For additional information, contact Michael Blumenthal, Executive Director, Scrap Tire Management Council, 1400 K Street, N.W., Suite 900, Washington, D.C. 20005.

CHAPTER **8**

Ground Rubber and Civil Engineering Markets for Scrap Tires

John R. Serumgard

INTRODUCTION

Each year more than 250,000,000 scrap tires are discarded in the U.S., representing about 3,000,000 tons of material, or slightly less than 2% of the municipal solid waste stream. Current public policy and social norms seek to reduce the potential harmful impact of waste materials, discourage simple disposal of waste materials and promote the sound reuse and recycling of as much waste material as is technically and economically possible. To apply these principals to the sound management of the 250,000,000 scrap tires generated each year in the U.S. requires the development of markets that can utilize the tires or the materials which can be developed from the scrap tires in beneficial ways.

Several markets for scrap tires do exist. The largest current market, and the market with the greatest short and intermediate range potential, is the utilization of the tire as a fuel. This market segment is the subject of a separate paper in this program, and thus will not be further discussed. In addition to the fuel market, there are several other markets currently in development that make use of the unique properties of the tire and the elastomeric material from which it is constructed. The subject of this paper is an overview of two of these major markets: the use of ground rubber as a raw material and the use of whole or processed tires in civil engineering applications.

GROUND RUBBER

Tires are manufactured from natural and synthetic rubbers compounded with a variety of processing aids, vulcanizing agents, anti-ozonants, and other mate-

1-56670-215-1/97/$0.00+$.50
© 1997 by CRC Press, Inc.

rials, reinforced with steel and fabric body plys, belting, and beads. From both engineering and chemical perspectives, tires are very sophisticated products. The most essential step in the manufacture of a tire is the vulcanization process, which combines all of the various tire components into a coherent whole and which transforms the rubber into a temperature stable material. During the vulcanization process, pressure and temperature combine to cause chemical and physical changes in the materials in the tire. As a thermoset material, the rubber in tires is permanently changed by the vulcanization process. Only limited success has ever been achieve at completely reversing these changes.[1]

One method of liberating the rubber from a tire so that it can be reused is to grind it into a crumb or powder while removing the steel and fabric components. Ground rubber is produced in one of two basic methods — ambient grinding and cryogenic grinding. Both start with tires being passed through shredders that reduce the tire to medium- to small-sized pieces. These smaller pieces are then fed through additional size reduction equipment such as cracker mills, or grinders for the ambient process. In the cryogenic process, the tire pieces are temperature reduced to the glass transition temperature and then ground. The most commonly used cryogenic processing agent is liquid nitrogen. In both processes, reinforcing steel and fiber materials are removed. The result in either case is a ground vulcanized rubber material, which can range in size from fairly coarse material in the 10 to 20 mesh range or larger, to much finer sizes in the 80 to 200 mesh or finer ranges.

Other waste rubber products, such as production scrap from the manufacture of molded and extruded products or from hose and belt manufacturing, or from any other rubber products manufacturing operations can also be reduced to ground rubber. A tire may contain several types of rubber, and different rubber compounds. As a result, the chemical properties of ground rubber produced from whole tires can vary. In addition, different manufacturers use different compounds, and compounding recipes differ with the type of tire. This adds to the variability possible with ground tire rubber. In contrast, production scrap or off-specification products from other, non-tire, rubber manufacturing processes may contain only a single rubber compound. When this material is ground, it may present more uniform characteristics and may more easily be utilized, including being used back in the manufacture of the original product from which it came.

There are several major uses for ground rubber. It is used in compounding rubber stocks for use in production of various molded rubber products. Ground rubber is also being used with higher cost elastomers to find ways to reduce material costs without affecting product performance. It can be used as a modifier for asphalt paving materials. Ground rubber has also been mixed with other materials, both recycled and virgin, to manufacture products not traditionally made from rubber. Experimentally, it has been used as the basis for material to remediate oil spills.

Ground rubber is used in the manufacture of many products. Major markets included:

- Friction materials
- Floor mats and relief pads
- New tire manufacture
- Molded rubber products

In new tire manufacture, ground rubber is most commonly used as a processing aid. Small quantities of ground butyl rubber are used in compounding the tire innerliner to reduce air entrapment during curing. In addition, small quantities of ground rubber are used as filler materials in various tire components where its use does not compromise the performance required of the specific elastomer compound.

Floor mats, carpet backing, and relief mats have been traditional markets for ground rubber where the material adds resiliency at a lower cost compared to virgin compounds. Certain molded products have also been markets for ground rubber. An interesting use is the manufacture of soaker hoses from ground rubber and a polyurethane binder.

The key to expanding product markets for additional recycled ground rubber is the acceptance of the material by customers. The auto industry is one of the largest customers for molded rubber products. Until recently, most auto industry specifications for molded rubber parts specifically prohibited the use of recycled rubber. Increasingly, the auto industry is now seeking greater use of recycled materials, including recycled rubber, in the parts ordered and purchased. An example is the recent announcement by Ford that it is testing the use of a compound containing ground recycled rubber to manufacture brake pedal pads. This is a welcome move to expand ground rubber markets.

Another need has been a technical forum in which rubber industry chemists and engineers can share information and insights into the problems raised by expanded use of recycled ground rubber. This need is being filled by the recent organization of a Topical Discussion Group on Rubber Recycling organized as part of the Rubber Division of the American Chemical Society. This group sponsored its first symposium this past Spring as part of the Rubber Division's semi-annual meeting. The Topical Discussion Group brings together the producers of ground recycled rubber, with compounding and engineering personnel from rubber products manufacturers and representatives from customer industries to share information and identify needed activity.

The morphology of ground rubber particles plays an important role in the ability to incorporate them into new compounds. Ambiently-ground particles tend to have a more varied surface than cryogenically-ground particles. The cryogenic process tends to produce particles with smoother surfaces that are more regular, cube-shaped pieces. In practice, the adhesion of ground rubber in new compounds is a mechanical process. As a result, ambiently-ground material appears to more easily incorporated into most compounds. In order to increase the ability of cryogenically-produced material to be incorporated in new compounds, work has been done to modify the surface of these particles to make them more reactive when incorporated into new compounds. In particular, Bauman has developed such a process, and has received assistance from the Department of Energy to move it to commercial applications. Given the higher cost associated with this surface modification process, a principal area of investigation is the use of this material in compounds using high value virgin elastomers.

ASPHALT RUBBER

Ground recycled rubber has been developed as a modifier for asphalt paving materials over the past 40 years in both the U.S. and in Europe. The original developmental work was undertaken in an effort to improve the performance of asphalt paving materials, not to find a method to dispose of scrap tires. In the search to improve asphalt paving materials, many different additives and modifiers have been tested, with many making the grade to become accepted practices. The utilization of various polymers has been one of those avenues of exploration.[2] Eventually, some attention was given to crumb rubber. Two basic processes have been developed — the "wet" process and the "dry" process. The difference is the point at which the ground rubber is introduced.

The wet process is also called the McDonald or Arizona process after its developer and the location in which it was developed. Ground rubber is prereacted with the hot liquid asphalt cement for a specified period of time, depending on the size of the rubber particles. During this process the rubber particles react with and change the asphalt cement. The rubber modified asphalt can then be used in one of several ways. It can be used as a crack sealant in pavement repair. Modified asphalt can be used with an aggregate chip as a cape sealant to effect temporary repairs. In the rehabilitation of cracked pavements to prevent or reduce subsequent reflexive cracking, it can be sprayed over cracked pavements as a stress absorbing membrane (SAM) or with a thin layer of asphalt concrete as a stress absorbing membrane interliner (SAMI) and then covered with a new friction course. Rubber-modified asphalt cement can also be mixed with the normal aggregate component to produce a rubber-modified asphalt concrete. This material can be placed in the usual manner, with some limited modifications, and used in both base and friction courses.

The dry process was originally developed in Sweden and involves using slightly larger sized rubber particles mixed with the dry aggregate materials. The aggregates, which must be gap graded to allow room for the rubber particles, are mixed with normal asphalt cement to form a modified asphalt concrete. Again, the modified material can be placed in the usual manner with limited modifications to the handling procedure. For more information see Collins et al.[3]

At present the largest single issue involved in the use of crumb rubber modifiers for asphalt paving is the mandate contained in Section 1038 of the Intermodal Surface Transportation Efficiency Act (ISTEA), passed by Congress in 1991. Under Section 1038, all states were required, beginning in 1994, to use crumb rubber derived from scrap tires in a specified portion of the state's federally aided highway construction. The mandate has been challenged in a major way by the asphalt paving community and by the state departments of transportation (DOTs). For its part, the Federal Highway Administration (FHWA) has undertaken to complete the studies required of it and to provide guidance to the various states which must comply. (See Crumb Rubber Modifier Workshop: Design Procedures and Construction Practices, Arlington, Texas, 1993, U.S. Department of Transportation, Federal Highway Administration.)

Ground rubber used as an asphalt modifier is a reasonably accepted material in only a few areas of the country, most notably Arizona, California, Florida, and Texas. Greater acceptance will require that the state highway departments establish to their own satisfaction that rubber modification of asphalt paving materials creates a material with positive attributes, and thus become confident about the material and its capabilities. In turn this may require some additional testing and additions to the current body of engineering knowledge about the material and its performance, especially in geographically distinct regions of the country. Efforts are currently underway to ensure greater technology transfer and information dissemination on asphalt rubber issues. The National Cooperative Highway Research Program of the National Research Council has established the CRM Project, Project 20-7, to provide an information network for anyone interested in crumb rubber modifiers. The project is administered by the Technology Transfer Center at the College of Engineering, University of Nevada, Reno. The same organization has published *Uses of Recycled Rubber Tires in Highways: A Synthesis of Highway Practice,* which contains over 500 references through 1993.

NON-TRADITIONAL PRODUCTS

In recent years there has been considerable interest in ground rubber as a raw material with many experimenters and entrepreneurs attempting to develop totally new uses for the material. Some of the more interesting ones have included the use of ground rubber and recycled plastics to produce a lumber substitute. Among the products produced in this manner have been fence posts and flooring for livestock trailers. Other investigators are experimenting with ground rubber as an additive to concrete to give greater resistance to earthquake damage, or to achieve other objectives. Also at the experimental level is the use of rubber powders in agents used to remediate oil spills.

STANDARDIZATION

All ground rubber markets will benefit from a greater level of standardization. To that end, the Scrap Tire Management Council is working with several ASTM committees to develop size standards for ground rubber materials in various application. Involved in this effort are ASTM Committees D-11.26, Rubber;

D-4.45, Modified Asphalt Specifications Pavements and D-34.15, Construction and Other Secondary Applications of Recovered Materials. With the development of specifications, both producers of ground rubber and their customers will benefit as a market can then more fully develop.

FURTHER MANUFACTURING

Broader use of ground rubber in rubber parts manufacturing will require greater collaboration between rubber parts producers and their suppliers and customers to work on the process of incorporating this material into their products. The key to this market will be the customers for the parts being produced. In this context, the auto industry will be the most important market as it consumes a large percentage of the non-tire rubber products production. The auto industry establishes rigid specifications for the parts it purchases, including, in the case of rubber parts — until recently — prohibitions against the use of recycled rubber. As the interest in recycling has expanded, so, too, has the auto industry interest in the recycling and recyclibility of its products. The industry has established a cooperative program in the field, the Vehicle Recycling Partnership (VRP). In addition, some auto manufacturers have recently announced specific projects that include recycled rubber. Ford is investigating the use of ground rubber in a compound for its brake pedal pads. Chrysler has announced that it is already using a rubber shield which incorporates crumb rubber derived from scrap tires. In each case, these companies also indicated a desire to expand the use of recycled post-consumer scrap rubber in its products.

Expansion of any use of recycled rubber in new products will require a close collaboration between the parts producer, the customer and the recycled material supplier. Reforumlation of rubber compounds will normally be required in order to incorporate significant amounts of recycled rubber without compromising the performance required of the finished part.

Information exchange is also an important requirement for expansion of this market segment. Normally, there is a high degree of proprietary knowledge in the rubber products industry. This has a tendency to limit dissemination of information. To counter this tendency, the Rubber Division of the American Chemical Society recently established a Topical Discussion Group on Rubber Recycling. The Group hopes to encourage presentations from companies that have successfully utilized crumb rubber in their products. The Group will organize seminars at regular meetings of the Division and will publish a newsletter.

CIVIL ENGINEERING

The civil engineering market for scrap tires encompasses several distinct uses. Whole tires have been used to construct retaining walls and crash barriers. One publicized use is the construction of houses and at least one motel using whole tires to contain rammed earth. Whole tires have been used to construct breakwaters and artificial reefs, and in erosion control.

The broader use of scrap tires as an engineering material is the use of chipped or shredded scrap tires as a fill material. Tires are especially useful where a lightweight fill material is indicated. Notable projects include the fill over an underground garage in Minneapolis and the use of shredded tire fill to construct a ramp to an Interstate highway where the ramp passed over a closed landfill. The lighter weight of the tire fill is expected to reduce settling as the landfill matures. Research is also being conducted in the use of shredded tires to backfill retaining walls, and as an insulating material under roadbeds in cold climates.

Expansion of the civil engineering markets depends on the availability of standard engineering data for shredded tire material. Fortunately, several of the projects are being undertaken by university based researchers who are calculating this data. See generally references by authors including Manion,[4] Upton,[5] Read,[6] Humphrey[7,8] and Ahmed.[9] It is likely only a matter of time before these data can be collected and published so that others can use them. Another limiting issue is a concern about the potential leachates that might result from any of these civil engineering uses. Again, a body of data is being developed that so far has demonstrated no difficulties with tires in these uses. See RMA[10] and Edil.[11]

Several state highway departments have undertaken road construction-related fill projects using tire chips. These include Oregon, Colorado, North Carolina, Virginia, Maine, Minnesota, and Wisconsin, among others. Minnesota has experienced considerable private use of tire chips as fill material, as has Vermont.[12] Recently several states have approved the use of tire chips in private septic systems as backfill for leach fields. These include Iowa, Georgia, and South Carolina.

Small tire chips have also been utilized as a soil amendment to improve athletic playing fields. A patented process marketed under the name Rebound™ combines crumb rubber from scrap tires with composted organic material to reduce soil compaction, resulting in better athletic playing surfaces. Installations have been made in Florida, California, Colorado, Hawaii, Maryland, Michigan, Missouri, Nevada, Virginia, and Wisconsin.

The Americans with Disabilities Act (ADA) has opened another market for scrap tires — the resilient playground surface market. Many schools and recreation facilities are having to reconfigure playgrounds to make them more accessible, including finding resilient surfaces which can be traversed by wheelchairs. Crumb rubber makes an excellent playground surfacing material, and several manufacturers have moved into this market.

The goal of most scrap tire utilization projects is to find markets for scrap tires so that they will not end up in landfills or on stockpiles. Ironically, one potentially significant use of tires is in the construction and management of landfills, including Subtitle D qualified facilities. Both shredded and whole scrap tires have been approved in various states for use in constructing leachate beds in landfills. Approval has also been given in some states for the use of shredded tire material as a partial replacement for required daily cover.

ENVIRONMENTAL CONCERNS

A significant concern in the use of scrap tires is the potential for environmental problems associated with various civil engineering applications, most notable the leaching characteristics of shredded tire material placed in the earth. The evidence available to date suggests no significant concerns for use under conditions normally encountered in the environment. See Blumenthal.[3]

CONCLUSIONS

Markets for sound use of scrap tires are expanding, from less than 11% of annual generation in 1990, to more than 37% currently. Greater use of scrap tires in selected civil engineering applications, and the greater acceptance of ground rubber in markets such as asphalt rubber and rubber products manufacturing will aid substantially in increasing the total use of scrap tires to 100% of annual generation.

REFERENCES

1. Warner, W. C., Methods of Devulcanization, *Rubber Chemistry and Technology*, 67, 559, 1994.
2. Lewandowski, L. H., Polymer Modification of Paving Asphalt Binders, *Rubber Chemistry and Technology*, 67, 447, 1994.
3. Collins, R. J. and Ciesielski, S. K., Highway Construction Use of Wastes and By-Products; Ramaswamy, S. D. and Aziz, M. A., Some Waste Materials in Road Construction; Ahmed, I. and Lovell, C. W., Use of Rubber Tires in Highway Construction; Blumenthal, M. and Zelibor, J. L., Scrap Tires Used in Rubber Modified Asphalt Pavement and Civil Engineering Applications, in *Utilization of Waste Materials in Civil Engineering Construction,* Inyang, H. I. and Bergeson, K. L., Eds., American Society of Civil Engineers, 1992.
4. Manion, W. and Humphrey, D., Use of Tire Chips as Lightweight and Conventional Embankment Fill, Maine Department of Transportation, 1992, 149.
5. Upton, R. and Machan, G., Use of Shredded Tires for Lightweight Fill, Paper presented at Transportation Research Board, Washington, D.C., 1993.
6. Read, J., Dodson, T., and Thomas, J., Experimental Project Use of Shredded Tires for Lightweight Fill. FHWA Experimental Project No. 1, DTFH-71-90-501-OR-11. Oregon Department of Transportation, 1991.
7. Humphrey, D., Sanford, T., Cribbs, M., and Manion, W., Shear Strength and Compressibility of Tire Chips for Use as Retaining Wall Backfill, Paper presented at Transportation Research Board, Washington, D.C., 1993.
8. Humphrey, D. and Manion, W., Properties of Tire Chips for Lightweight Fill, in Grouting, Soil Improvement and Geosynthetics, Volume 2, Geotechnical Special Publication No. 30, American Society of Civil Engineers, 1992, 1344.
9. Ahmed, I. and Lovell, C., Rubber Soils as Lightweight Geomaterials, paper presented at Transportation Research Board, 1993.

10. Rubber Manufacturers Association (RMA), A Report on the RMA TCLP Assessment. Radian Corporation, Austin, Texas, 1990, 38.
11. Edil, T. B., Bosscher, P. J., and Eldin, N. N., Development of Engineering Criteria for Shredded or Whole Tires in Highway Applications, Department of Civil and Environmental Engineering, University of Wisconsin-Madison, WI, 1990, 19.
12. Frascois, R. I., Use of Tire Chips in a Georgia, Vermont Town Highway Base, Research Update No. U91-06. Vermont Agency of Transportation, April 17, 1991, 5.

CHAPTER 9

The State-of-the-Art in Municipal Solid Waste Combustion in the United States

Jonathan V. L. Kiser

INTRODUCTION

Communities across the country are increasingly recognizing that solving their trash problems requires a combination of modern management options. Sometimes this includes state-of-the-art waste-to-energy (WTE) facilities. These plants burn trash at very high temperatures to reduce its volume by up to 90%. State-of-the-art facilities are equipped with the most advanced air pollution control and monitoring equipment, and are operated by certified professionals who employ good combustion practice and ensure that ash residues are properly managed. Waste-to-energy plants also generate clean, renewable energy and are typically linked to recycling efforts, both in the community and on-site.

INDUSTRY STATUS

A total of 114 waste-to-energy plants now operate in the U.S. (Table 9.1). These facilities have a rated municipal solid waste processing capacity of nearly 101,000 tons per day (tpd) and process 15% of the U.S. municipal waste stream. About 38,900,000 people are currently served by these facilities in more than 1,600 communities in 32 states (Table 9.2). As of May 1996, there were 5 waste-to-energy facilities under construction in the U.S. with a rated design capacity of 2,645 tpd. An estimated 37 WTE facilities now are in the advanced planning stage with rated design capacity of about 10,500 tpd. If all of these facilities become operational, they will increase the quantity of waste handled at WTE plants to 16% of the 218,000,000 tons forecasted by U.S. EPA to be generated in the U.S. by the turn of the century.

1-56670-215-1/97/$0.00+$.50
© 1997 by CRC Press, Inc.

Table 9.1   Status of Waste-to-Energy Plants in U.S.

Technology	Number of operating plants	Daily design capacity (tons per day)	Annual capacity (million tons)[1]
Modular	21	2,529	0.8
Mass burn	70	72,813	22.6
RDF	23	25,579	7.9
Total WTE facilities	114	100,921	31.3

[1] Annual total assumes that plants operate at 85% of design capacity.

From *1996 IWSA MWC Directory,* Integrated Waste Services Association, Washington, D.C., 1996. With permission.

Waste-to-energy plants are divided into three primary technology types:

1. Mass burn plants, which burn mixed municipal waste typically in a single combustion chamber surrounded by water-filled tubes for heat recovery, and are constructed in the field;
2. Refuse-derived fuel (RDF) plants, which first remove recyclables and unburnable materials and process the remainder into a uniform fuel. The fuel is typically burned in a dedicated boiler on site; and
3. Modular facilities, which are similar to mass burn plants but are typically smaller in size, have two-chamber combustion, are manufactured in a shop, and are erected at the designated site.

Table 9.2   Key Facts Pertaining to Operating Waste-to-Energy Facilities

Percentage of U.S. waste managed by WTE	15%
Population served[1]	38.9 million
Communities served[2]	1,631
Number of states with plants	32
Average community recycling rate[3]	26%
On-site recycling[4]	856,000 tpy
Number of homes supplied with electricity[5]	1.3 million

[1] Based on a municipal waste generation rate of 4.4 lbs/person/day.
[2] Based on the response from representatives of 106 WTE projects.
[3] Based on the response from representatives of 74 projects.
[4] Based on the response from representatives of 93 WTE projects. Includes 739,000 tons of ferrous metals and 117,000 tons of other recyclables.
[5] Based on a net WTE facility electricity export equivalent basis.

From *1996 IWSA MWC Directory,* Integrated Waste Services Association, Washington, D.C., 1996; and Kiser, J. V. L., Calculations, 1996. With permission.

AIR POLLUTION CONTROL

Modern waste-to-energy plants employ Good Combustion Practice (GCP) as the basis for ensuring proper emission control. GCP is used to minimize products of incomplete combustion, including carbon monoxide and organics. Proper furnace design accomplishes GCP by allowing for optimal furnace temperature, residence time, and turbulence.

In addition to GCP, acid gas scrubber/fabric filter combinations and electrostatic precipitators (ESP) are the two most common air pollution control systems reported to be in place at existing operations (see Table 9.3). Fifty-five percent of the waste-to-energy plants are equipped with a combination scrubber/fabric filter or scrubber/ESP; 30% are equipped with fabric filter or an ESP only; and 16% have nitrogen oxide control. Separate from the air pollution control equipment, 75% of the facilities also reported having continuous emissions monitoring equipment, and 17% are directly linked via computer to the regulatory enforcement agency to enable 24-hour monitoring of the continuous emissions monitoring printouts.

For those waste-to-energy projects now under construction and being planned, there is an ongoing trend toward the use of a scrubber/fabric filter combination, along with NOx and mercury control using carbon, plus continuous emissions monitors, and computer linkage to the regulatory agency (Table 9.4). This trend will be ensured by the recently promulgated Maximum Achievable Control Technology (MACT) requirements under the 1990 Clean Air Act Amendments (CAAA). Bearing these CAAA in mind, many of the Table 9.3 statistics will change to reflect the trend toward more controls and monitors. In other words, waste-to-energy plants are expected to be equipped with scrubber/fabric filters, will address NOx and mercury control in some fashion, and have continuous emissions monitors.

CAAA MACT REGULATIONS

EPA's regulatory update to the 1991 New Source Performance Standards (NSPS) and Emission Guidelines (EG) was proposed on September 1, 1994. The update, in the form of MACT regulations for municipal waste combustors, was originally scheduled for release in 1991. The 1994 date resulted from a consent decree issued by the U.S. District Court in litigation by the Natural Resources Defense Council and Sierra Club against the EPA. The Integrated Waste Services Association also sued the Agency in November 1993 over missed deadlines to promote MACT resolution.

The MACT package was submitted to OMB in early June and signed, per the consent order, by EPA Administrator Carol Browner in September 1, 1994. The proposed standards were published in the *Federal Register* on September 20, 1994. The promulgated standards were published in the *Federal Register* on December 19, 1995.

The MACT regulations establish, for the first time, national numerical emissions limitations for lead, cadmium, and mercury. Dioxin emission limits have also been included for new and existing plants.

Table 9.3   Type of Air Pollution Control and Monitoring Equipment for Existing Waste-to-Energy Plants[1,2]

Plant type	Scrubber/FF (%)	Scrubber/ESP (%)	Percentage of equipment						
			ESP (%)	FF (%)	CARBON (%)	LIME (%)	NOx (%)	CEM (%)	LINK (%)
Modular	14	14	38	10	—	14	—	57	5
Mass burn	54	13	30	4	16	27	21	87	23
RDF	26	17	17	9	9	22	13	52	2
Total WTE	41	14	24	6	11	24	16	75	17

[1] Reflected in Table 9.3 are the types of air pollution control and monitoring equipment now operating at WTE plants. The numbers represent the percentage of each equipment type found at each type of facility.

[2] The number of facilities reported here includes 21 modular plants, 70 mass burn facilities, and 23 refuse-derived fuel plants, for a combined total of 114 waste-to-energy facilities.

From *1996 IWSA MWC Directory*, Integrated Waste Services Association, Washington, D.C., 1996. With permission.

Table 9.4　Type of Air Pollution Control and Monitoring Equipment for Waste-to-Energy Plants Planned and Under Construction[1,2]

Plant type	Scrubber/FF (%)	Scrubber/ESP (%)	FF (%)	Percentage of equipment CARBON (%)	LIME (%)	NOx (%)	CEM (%)	LINK (%)
Modular	—	—	100	—	—	—	100	—
Mass burn	100	—	—	80	40	100	100	60
RDF	100	20	—	80	60	100	100	60
Total WTE	91	9	9	73	45	91	100	55

[1] Reflected in Table 9.4 are the types of air pollution control and monitoring equipment reported for the WTE plants planned and under construction. The numbers represent the percentage of each equipment type reported for each type of facility.

[2] The number of facilities reported here includes 5 mass burn facilities, 5 refuse-derived fuel plants, and 1 modular plant for a combined total of 11 waste-to-energy facilities.

From *1996 IWSA MWC Directory*, Integrated Waste Services Association, Washington, D.C., 1996. With permission.

In terms of how much time existing plants have to retrofit to comply with the new standards, facilities have up to five years from the promulgation date, depending upon local conditions and need.

OPERATOR CERTIFICATION

The 1991 NSPS and EG contains a provisional certification standard for waste-to-energy operators across the country. This operator certification is required for all plant managers and shift supervisors. Three parameters are included in the standard: education (in the form of a relevant degree); time (i.e., years of experience operating power plants); and the passing of the certification test.

EPA stipulated that waste-to-energy operators be certified by February 1993. Contained within the new MACT provisions are full certification requirements. In addition to a written exam set forth under provisional certification, the full certification program calls for two representatives from the American Society of Mechanical Engineers (ASME) and one representative from a regulatory agency to visit each plant. These representatives will review how current plant operating manuals are, plus conduct an oral exam that tests the operator's working knowledge of plant operating procedures.[1]

POWER GENERATION

Burning municipal solid waste at waste-to-energy plants results in both significant volume reduction and offers the potential for energy recovery. Energy generation occurs through the production of steam in a boiler which, in turn, can be put to beneficial use. For example, steam or hot water generated by the process can be used for space heating in buildings, as process heat for industrial operations, or for district heating/cooling system applications. Steam also may run a turbine and generate electricity. The type of power that is ultimately produced at a waste-to-energy plant depends upon the markets the facility serves.

Under federal law, electric utilities are required to purchase power from qualifying facilities like waste-to-energy. Industrial markets for steam are often more economically attractive, depending upon the internal cost of steam product, but are also often more difficult to locate. Regardless of the ultimate market, the price of power from waste-to-energy plants is negotiable. Successful negotiation of a power contract is critical to the economic viability of most projects. Revenues from energy sales help to offset facility capital, operations, and maintenance costs.

The power generation capability of a waste-to-energy facility is typically based on the overall heat release rate of the waste burned (e.g., million Btu/hour), not the mass throughput of the plant (e.g., tons per day). As the waste heating value (e.g., Btu/lb) changes due to variations in waste composition, the mass throughput of the plant will fluctuate accordingly. For example, a waste-to-energy plant designed to process waste with an average heating value of 5,000 Btu/lb would require a reduced throughput to compensate for an increase in the waste heating value due to the removal of glass, metals, high moisture yard waste, and other materials separated for recycling.[2]

THE VALUE OF WASTE-ENERGY

Energy production is not the primary reason why communities turn to waste-to-energy plants to help meet their needs — protecting the public health through reliable, long-term waste disposal is most important. The generation of energy nonetheless provides important economic and environmental benefits. In addition to providing a domestic source of renewable energy, thereby reducing the need for fossil fuels, the production of energy reduces the cost of facility operations. Revenues from the sale of power, coupled with the tipping fee (i.e., the cost charged to process waste at the facility), help to pay the debt service (i.e., the facility's mortgage), and the operations and maintenance costs. Other revenues increasingly result from the sale of recyclable materials recovered on site prior to and after combustion, interest on reserve funds, and other sources. Energy sales alone generally account for 35 to 50% of a facility's revenues.

WASTE-TO-ENERGY POWER EQUIVALENTS

Operating waste-to-energy facilities have an electricity generating capacity equivalent to approximately 2,300 megawatts (on a conservative, gross basis). This translates into about 17.1 million megawatt hours of exportable energy per year to market.

Table 9.2 illustrates that the net power generating potential of waste-to-energy operations is equivalent to meeting the electricity needs of about 1.3 million homes. Viewed in other terms, the facilities generate enough exportable energy to supply electricity to about 430 Empire State Buildings, or the annual equivalent of displacing 31 million barrels of foreign oil (using the conversion of crude oil into electrical energy as the basis).

ASH MANAGEMENT

In September 1992, former EPA Administrator William K. Reilly issued a memorandum to all EPA Regional Administrators declaring the ash from non-hazardous municipal waste combustion is exempt from hazardous waste regulation under Subtitle C of the Resource Conservation and Recovery Act (RCRA). The Administrator pointed out that the statutory goals in Section 3001(I) of RCRA — environmental protection and the promotion of resource recovery — were best served through this declaration. Further, ash regulation under EPA's 1991 municipal landfill criteria was sufficient to ensure full protection of human health and the environment.[3] EPA's position was grounded in the growing body of empirical field data showing that MWC ash can be safely managed with little difficulty.

In spite of the Reilly memorandum and ample empirical field data, on May 2, 1994, the Supreme Court ruled in *City of Chicago vs. Environmental Defense Fund* [114 S.Ct. 1588, (1994)] that ash is not exempt from Subtitle C regulations under RCRA. The Court's decision requires that ash exhibiting hazardous characteristics be treated as hazardous waste. If the ash does *not* exhibit such characteristics, it may continue to be disposed in municipal waste landfills.

The EPA implemented the Supreme Court's decision three weeks later. The Agency announced, in an effort to be wholly consistent with the Court, that ash exhibiting hazardous characteristics must be managed accordingly on June 1, 1994, the effective date of the ruling. The Agency also issued an implementation strategy on May 27th, draft ash sampling guidance, and a *Federal Register* notice seeking comments by September 21, 1994 on its draft guidance document.

In anticipation of Congressional action on ash management, a hearing was held on July 27, 1994 by the House Energy and Commerce Subcommittee on Transportation and Hazardous Materials. In early September, draft legislative language was subsequently submitted on Capitol Hill by a diverse coalition including Environmental Defense Fund, Integrated Waste Services Association, National Association of Counties, National League of Cities, Solid Waste Association of North America, and WMX Technologies, Inc. This proposed bill called for ash to be regulated as a municipal, not hazardous, waste. Ash would be disposed in monofills meeting strict design and operation standards, and limited ash utilization applications would be permitted. The proposed bill was not acted upon by the 103rd Congress.

In the spring of 1995, EPA issued its ash sampling guidance that provided important clarification pertaining to the treatment of ash. The Agency stipulated that ash could be treated on site within "the four walls" of a waste-to-energy facility before it would have to be subjected to any toxicity testing requirements.

RECYCLING COMPATIBILITY

A survey was conducted in 1992 to determine the compatibility of waste-to-energy and recycling. Sixty-six communities were contacted, with 51 currently having a waste-to-energy plant, two with a facility under construction, and 13 planning to implement waste-to-energy.[4]

The survey respondents included waste management professionals in each community, including municipal officials, recycling coordinators, waste management authority personnel, industry representatives, a consultant, and a trade association employee. Each person was asked questions regarding the type of existing recycling programs, whether the recycling program(s) was expanding, the current recycling rate, what community recycling program(s) really stand out, and to provide supporting evidence regarding whether or not waste-to-energy and materials recycling are compatible.

The survey results suggest that materials recycling and waste-to-energy work better together than they do apart. Primary survey findings included:

- A majority of the communities serviced and planning to be serviced by waste-to-energy have recycling rates exceeding the national average — some nearly three times.
- Recycling programs are expanding in all of the responding communities.
- 92% of the waste-to-energy projects have/will have some form of onsite recycling.
- All of the waste-to-energy facility projects are linked to offsite community recycling programs.
- Every municipal official, recycling coordinator, and other waste management professional contacted said waste-to-energy and materials recycling are compatible, and provided supporting evidence.

Follow-up surveys were conducted in the fall of 1993 and in the spring of 1995 and 1996. The results from the 1993 effort revealed that the average recycling rate for communities with waste-to-energy plants was 23%. By the spring of 1996, 74 communities across the U.S. with operating waste-to-energy facilities had an average recycling rate of 26%.

In addition, a combined total of about 739,000 annual tons of ferrous metals are recovered onsite for recycling at 79 waste-to-energy projects (see Table 9.2). Most of these materials were "automatically" recovered post-combustion by way of magnets. 117,000 tons of non-ferrous and other materials are likewise being recovered through front-end processing and post-combustion eddy current separation technology.

Many of these recycled materials would not otherwise have been recovered through municipal curbside, drop-off, or other recycling programs. One reason is because they are not specifically included in community recycling programs. Table 9.5 helps to further illustrate this point by listing some of the types of materials recovered during a 1992/1993 post-combustion metals separation demonstration project at the Fairfax County, VA plant.

HEALTH AND ENVIRONMENTAL SAFETY

In the summer of 1993, public officials, independent engineers, industry environmental experts, risk communicators, and corporate executives teamed up to investigate the health and environmental safety of waste-to-energy. The result of this year-long effort, which required extensive technical input, over 40 reviewers, a dozen drafts, and at least 1,000 comments, is a public information booklet called "Clean, Renewable, Safe and Economical America's Newest Energy Source."[5]

A primary finding of this investigation was that, compared to other fuels burned to generate steam and electricity, modern waste-to-energy facilities are very clean. In fact, waste-to-energy plants typically *improve* air quality by replacing fuels like oil and coal burned at utility power plants. Utilities typically take their older fossil fuel capacity off-line when purchasing waste-to-energy power.

Table 9.5 Select Materials Recovered Post-Combustion at the Fairfax, VA Waste-to-Energy Plant

Ferrous metals	Nonferrous metals
Can opener hardware	Salt and pepper shaker tops
Toasters	Pots and pans
Eating utensils	Garden hose ends
Flashlight cases	Lawnmower frames
Roller skates	Power cords
Bicycles	Wind chimes
Ironing boards	Canoe paddles
Luggage hardware	Lawn sprinklers
Baby strollers	Bullet casings
Belt buckles	Door locksets
Hair dryers	Clock internals
Lamps and light fixtures	Candle holders
Closet shelving	Lightbulb bases
Exhaust mufflers	Hub caps
Wheel rims	Auto interior trim

Adapted from Hauck, P., Ogden Martin Systems, Inc., Tampa, FL, 1993.

Another study finding was that waste-to-energy plants reduce the buildup of carbon dioxide in the atmosphere by replacing fossil fuels. In so doing, waste-to-energy plants do not contribute to the buildup of greenhouse gases.

SUMMARY

The combination of stringent federal air regulations and ash management trends will further ensure that modern waste-to-energy plants are fully protective of human health and the environment. At the same time, waste-to-energy operations will continue to generate clean power and thereby help reduce our demand for fossil fuels. Finally, waste-to-energy will also continue to contribute to integrated waste management solutions across the country. This is evidenced by the many examples demonstrating how well waste-to-energy works together with recycling.

REFERENCES

1. Licata, A., personal communication, February 1996.
2. Kiser, J. V. L. and Burton, B. K., Energy From Municipal Waste: Picking Up Where Recycling Leaves Off, *Waste Age*, 39, November 1992.
3. Reilly, W. K., Memorandum To All Regional Administrators Regarding Exemption For Municipal Waste Combustion Ash From Hazardous Waste Regulation Under RCRA Section 3001(I), U.S. Environmental Protection Agency, Washington, D.C., September 18, 1992.

4. Kiser, J. V. L., Recycling and WTE: Working Well Together, *Solid Waste & Power*, 1993 Industry Sourcebook, 12, HCI Publications, Overland Park, KS, December 1992.

5. American Society of Mechanical Engineers, Integrated Waste Services Association, Municipal Waste Management Association, U.S. Conference of Mayors, Solid Waste Association of North America, The AIMS Coalition, *Clean, Renewable, Safe and Economical, America's Newest Energy Source,* 1994.

CHAPTER 10

The Role of a Municipal Solid Waste Incinerator in a Small Community

Karen B. Morrison, Robert W. Hartley, and Marc A. R. Cone

INTRODUCTION

The purpose of this chapter is to present the efforts of a small town in the handling and disposing of municipal solid waste (MSW) in an environmentally safe and efficient manner. In Maine, many of the landfills are nearing their capacity or being closed due to the inability to operate in compliance with regulations and requirements. This is forcing communities to find alternative methods of disposal.

The Strawberry Creek Recycling Center in the town of Harpswell opened in the spring of 1979, replacing two open burning dumps. Harpswell currently operates a MSW incinerator in conjunction with recycling and landfill programs. Harpswell was one of the first 15 municipalities in the state of Maine to have both mandatory recycling and a solid waste ordinance.

RECYCLING

The recycling program currently removes a number of items from the waste stream in order to reduce and improve air emissions, and reuse certain materials for future production. The items that are recycled include: metal cans, glass, corrugated cardboard, newsprint, scrap metal, and automobile batteries. White goods (large household appliances) are stored on site, and picked up periodically by a firm which recycles them. Maine Department of Environmental Protection (DEP) will also be requiring the recycling of household batteries. Batteries are a primary source of lead, acids, mercury, cadmium, and other metal residue in emissions and ash. To make this program successful, the citizens must be aware of the program and willing to take an active part by presorting their waste. A condition has been included in Harpswell's air emission license to require the town to provide the public with general information about the incinerator, recycling and solid waste programs, and instructions on specific items to be separated from the incinerator waste stream. The recycling center has separate bins in which citizens place the items that have been separated. Used motor oil is used by local businesses that have the capability to burn waste oil. Tires are also collected, and are picked up for use as a fuel by others.

SOLID WASTE

The solid waste program diverts items that are neither recyclable nor combustible from the waste stream. These items include; brown goods (electronics), construction and demolition debris, furniture, and yard waste. Clean wood and brush are currently open-burned but will soon be processed through a chipper and used as mulch in the landfill. This separation process produces reductions in the quantity of waste for incineration, thus decreasing and improving air emissions.

INCINERATOR DESCRIPTION

Harpswell operates a small (Kelley Model 1280) MSW incinerator rated at 1500 lbs/h. For the air emission license renewal, the town of Harpswell proposes to incinerate approximately 12 tons/day of waste. Waste is fed into the incinerator by the use of a hydraulic feeder. The feeder hopper is loaded with waste and the feeder door is closed. The vertical charging door on the incinerator is opened hydraulically and the hydraulic charging ram pushes the waste into the pyrolysis chamber. When the charging ram is retracted, the vertical charging door closes, and the feeder door is re-opened to accept the next load of waste.

The waste is ignited in the pyrolysis chamber by a propane-fired burner with heat input of 100,000 Btu/h. The pyrolysis chamber is designed to operate in a pyrolytic (starved air) condition. A blower supplies under fire air through distribution ducts that are embedded in the refractory. The blower supplies only a portion of the oxygen required for complete combustion. Water is sprayed into the chamber through a spray nozzle when combustion temperatures exceed desired levels, which are in the 1300–1400°F range.

A hydraulic cylindrical ash ram is used to periodically push the waste pile toward the back of the incinerator. This action serves two purposes: it breaks up the waste pile so that more waste is exposed to the heat, and it moves waste and ash away from the charging door to make room for subsequent charges. The ash that collects in the pyrolysis chamber during each operating day is removed from the unit prior to the start of the next operating day.

The combustion gases produced in the pyrolysis chamber flow to the thermal reactor, or "afterburner" section, which contains three propane-fired burners with a total heat input of 4,200,000 Btu/h and an inspirator. The inspirator is essentially a perforated metal shell through which combustion air is induced. The combustion gases flow out of the thermal reactor and into the refractory-lined stack.

LICENSING ISSUES

Initial Process

Harpswell called a pre-application meeting to discuss the renewal process. During this meeting, Maine DEP and Harpswell reviewed the requirements for renewal of their license and inspected the facility for familiarity of the process. Harpswell submitted the application and necessary documents as required and a consultant completed a Best Practical Treatment (BPT) analysis for the facility. From this information, a determination of licensing requirements was made, and the renewal was drafted.

Best Practical Treatment

Maine DEP regulations require a level of emission control from the incinerator that would represent BPT. This is reviewed at each license renewal. BPT for existing emissions equipment means the method that controls or reduces emissions to the lowest possible level considering: the existing state of technology; the effectiveness of available alternatives for reducing emissions from the source being considered; and the economic feasibility for the type of establishment involved. Levels of control vary by the type of source, as well as the age of the equipment and the location of the facility. A BPT analysis was submitted for the Harpswell's MSW incinerator. This analysis proposed the following:

- Replace the existing thermal reactor (afterburner) with a new Cleaver Brooks secondary chamber designed to provide a minimum of 1.0 second retention time for 1800°F gases. Secondary chamber will contain one modulating propane-fired burner and one modulating blower.
- Install a modulating barometric damper in the stack. This will maintain a slight negative draft within the stack (–0.05 to –0.15 inches of water).
- Replace existing control system with a state-of-the-art control system. This new system will control all incinerator functions, including feed cycle, ash ram, burners, blowers, damper, and water injection. The new controls will contain automatic lockouts that prevent waste from being charged unless temperature of secondary chamber is at or above 1800°F.
- Install a continuous chart recorder to record the combustion temperature in the primary and secondary chambers.

The blower that is currently supplying underfire air to the pyrolysis (primary) chamber will continue to serve this purpose. The blower and the burner currently serving the pyrolysis chamber will be tied into the new control system.

Along with the equipment upgrades outlined above, the operator will be provided with training on the proper use of the new control system, and the town will be provided with copies of the manufacturer's recommended operating procedures. To maintain the incinerator in proper condition, the consultant will continue the quarterly preventative maintenance inspections that are currently required by the town's existing air emission license.

The BPT analysis reviewed emission control methods for the following pollutants:

- Particulate Matter
- Carbon Monoxide and Volatile Organic Compounds
- Heavy Metals
- Acid Gases
- Chlorinated Organic Compounds
- Nitrogen Oxides
- Fugitive Particulate from Ash Handling

In summary, proper combustion from the new controls system, achieving the required retention time and temperature in the secondary chamber, proper operation and maintenance practices, and continuing the waste recycling and collections programs will minimize emissions from the incinerator sufficiently to make continued operation of the incinerator economically and environmentally feasible.

Licensing Conclusions

After review of the BPT analysis for the incinerator, and the recycling and solid waste programs in the town of Harpswell, the EPA determined that proper combustion control by the new control system, achievement of the required retention time and temperature in the secondary chamber, proper operation and maintenance practices, and the continuation of the waste recycling and collection programs will minimize emissions from the incinerator.

The incinerator will be limited to a charging rate of 1500 lb/h, for a maximum of 16 h/day; this equates to 12 tons/day because the final 4 h period is used as a burn off cycle. At this maximum rate of charge, the emissions from this source will be limited to the following:

| Pollutant | Emission Limits | | |
	Actual lbs/h	Maximum lbs/h	Maximum tons/yr
PM	2.0	2.7	7.9
PM_{10}	2.0	2.7	7.9
SO_2	5.0	18.0	18.4
NO_x	4.0	7.0	20.4
CO	2.6	7.5	21.9
VOC	0.1	2.3	6.7
Pb	0.1	0.2	0.6
HCl	5.4	17.0	20.4

In addition, the source will be limited to a particulate matter (PM) emission rate of 0.10 gr/dscf, corrected to 7% O_2, and an opacity limit of 20%, excluding the emission of water vapor.

All upgrades and modifications proposed in the BPT analysis were required to be in place by March 1, 1994, with compliance testing to be done by June 1, 1994, and the results submitted within 30 days of completion of such tests.

Stack Testing

As a requirement of the licensing process, Harpswell ran tests of the incinerator emissions to determine PM and carbon monoxide (CO) emission rates (before the application of the upgraded modifications). For PM, compliance (as determined by U.S. EPA reference method 5 testing) was marginal, at best. Two of the three runs showed results of 0.18 gr/dscf, below the DEP license allowed rate of 0.20 gr/dscf (corrected to 12% CO_2). The third run, in which there was an upset in the feeding of the MSW into the incinerator, resulted in an emission rate of 0.32 gr/dscf. Using a time weighted average, this would equate to 0.1975 gr/dscf. Given the error inherent in stack testing, this would be insufficient evidence to determine the status of compliance. However, with the proposed upgrades to the incinerator, emissions are expected to be even less than those already recorded; thus, there should be no problem attaining compliance. Testing was done to determine the emission rate for carbon monoxide; the results were 0.60 lb/h for the first run, with the two subsequent runs registering as nondetectable. These compare to the town permitted rate of 0.60 lb/h.

Stack testing was performed subsequent to the modifications, on May 10 and 11, 1994. Results of the testing determined a particulate matter emission rate of 0.087 gr/dscf (corrected to 7% O_2). Carbon monoxide emissions of 2 ppm (which equates to 0.03 lb/h) were measured. In both cases, an improvement over testing done prior to the modifications has been demonstrated.

ALTERNATIVES TO AN INCINERATOR

The town of Harpswell completed an economic comparison between the continued operation of the incinerator and transferring the waste offsite to decide the future of the facility. The cost to continue operation of the incinerator included: operating expenses covering labor and benefits for one person, propane, utilities, maintenance, vehicle fuel, cost of transfer of 200 tons of ash to a landfill, and capital costs for the incinerator portion of the original bond, the recycling center reserve bond, and the new bond for the incinerator upgrades.

The cost of transferring waste offsite included: operating expenses covering labor and benefits for one person, utilities, maintenance, vehicle fuel, tipping fees and trucking costs, equipment leasing, and capital costs for the incinerator portion

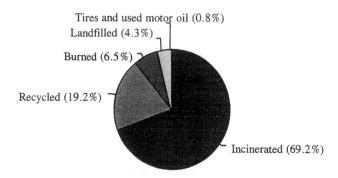

Tires and used motor oil (0.8%)
Landfilled (4.3%)
Burned (6.5%)
Recycled (19.2%)
Incinerated (69.2%)

Figure 10.1 Method of waste disposal for the town of Harpswell, ME; 1992 figures.

of original bond and the recycling center reserve bond. The final costs were estimated at $78/ton to operate the incinerator and $123/ton to transfer waste offsite. Transferring waste off site included tipping fees and trucking costs alone of $79/ton, which was greater than the overall cost to continue operation of the incinerator. In summary, even with the proposed upgrades to the incinerator, continued operation was more cost effective, saving approximately $45/ton. Figure 10.1 shows the breakdown of the disposal of the solid waste collected by the Strawberry Creek facility. The percentages are based on the annual collection of 2295 tons of MSW for the year of 1992.

CONCLUSIONS

The incineration and recycling programs at Strawberry Creek Recycling Center greatly reduce the volume of material that needs to be landfilled. By operating effective recycling and collection programs, Harpswell minimizes the emissions of metals and acid gases, which are prime concerns associated with MSW incinerators. The temperatures achieved in this unit are not conducive to the formation of dioxins and furans, which is another major concern. Lastly, proper operation and maintenance of the upgraded incinerator will minimize emissions of the criteria pollutants including PM, CO, VOC, and NO_x.

In conclusion, this program does not solve the solid waste problem, but it does greatly reduce the waste that is landfilled. Approximately 70% of the material collected by the Strawberry Creek Recycling Center is incinerated, with only 4% being landfilled. Although there are air emissions associated with the incinerator, these are controlled to an acceptable level by the separation of certain items from the waste stream, and proper operation and maintenance of the incinerator.

REFERENCES

1. MacMillan, C. J., *Best Practical Treatment Analysis, MSW Incinerator; Town of Harpswell, ME*, NMC Environmental Group, S. Portland, ME, December 1992.

2. Woodman, D. E., *Emission Test Report, Strawberry Creek Recycling Center Incinerator, Harpswell, ME*, Woodman Engineering, Inc., Tolland, CT, September 1992.
3. Henning, B. A., *Emissions Testing at Strawberry Creek Municipal Solid Waste Incinerator, Harpswell, Maine*, Air Pollution Characterization and Control, Ltd., Wayland, ME, May 1994.

CHAPTER 11

Air Emissions of Mercury and Dioxin: Sources and Human Exposure

Curtis C. Travis and Bonnie P. Blaylock

INTRODUCTION

Mercury is a widely used metal that can be detected in almost all environmental media. Anthropogenic sources of mercury include coal-fired power plants, municipal waste incinerators, burning of fuel oil, smelting and refining, and industrial sources such as chloralkali plants, electrical, chemical, and paint industries.[1,2] Natural sources of mercury include volatilization from soils, vegetation, and the ocean. Mercury, although extremely useful, is also very toxic and even fatal to humans, mammals, birds, and most other organisms. Clinical and subclinical effects have been found in occupationally exposed workers,[3] and several incidences of human fatalities have also been reported as a result of nonoccupational exposure to mercury. The most toxic form of mercury to humans is methylmercury. Because most chemical forms of mercury can be bacterially methylated in the natural environment, all forms of environmental mercury are a potential threat to humans.

2,3,7,8–Tetrachlorodibenzo-p-dioxin (TCDD) is the most potent chemical carcinogen ever evaluated by the U.S. EPA,[4] and a widespread fear persists that exposure to small amounts of TCDD could lead to serious adverse health effects. Anthropogenic sources of dioxin include municipal solid waste (MSW) incinerators, pulp and paper mills, motor vehicles, and residential wood burning. Water, soil, sediment, air, plants, animals, and humans worldwide almost always show detectable concentrations of mercury and dioxin. The fact that mercury and dioxin are ubiquitous in global environments[3,5,6] indicates that all humans are continuously exposed to quantifiable levels of these chemicals.

This chapter evaluates anthropogenic sources of mercury and dioxin, surveys background concentrations, and estimates the extent and pathways of human exposure to these chemicals in the U.S. for the purpose of determining the greatest sources of mercury and dioxin in the environment, and proposing best estimates of background concentrations. In addition, the authors estimate average daily intake using a terrestrial food chain model, described in detail elsewhere.[7]

MERCURY

Anthropogenic Sources

Estimates of national anthropogenic mercury emissions vary considerably in the literature. Recent estimates suggest that the total atmospheric emissions in the U.S. are about 650 tonnes/yr.[2] Our estimate is closer to about 790 tonnes/yr. The most important sources of mercury include commercial and consumer products such as paint and electrical equipment, fossil fuel combustion, smelting, and municipal waste incineration. These sources and estimates of their mercury emissions to environmental media are discussed below.

Commercial and Consumer Products

Industrial processing of commercial products and consumer use of mercury-containing products are two of the largest sources of mercury emissions. Industrial emissions of mercury result from volatilization during processing of products. Industrial sources of mercury include high-temperature processes in steel and iron manufacturing, cement production, mercury-cell chlor-alkali plants, industrial application of the metal, battery manufacturing, and laboratory and dental preparations. Watson[8] attributes close to 380 tonnes/yr of mercury emissions to commercial sources, accounting for about 48% of the total anthropogenic mercury emissions in the U.S. This estimate is lower than EPA's[9] estimate of 484 tonnes/yr of atmospheric emissions, or about 61% of the total. EPA's estimate includes emissions from chlorine and caustic soda manufacture (152 tonnes/yr), batteries (100 tonnes/yr), mildew proofing of paints (77 tonnes/yr), wiring devices (56 tonnes/yr), measuring and control instruments (34 tonnes/yr), and other less significant (<15 tonnes/yr) sources. Emissions from paints, in which mercury is used as a mildew retardant, contribute significant amounts of mercury to the environment. In 1983, the estimated mercury content of paint was 209 tonnes, and assuming a volatilization factor of 65%, the resulting emissions estimate was 136 tonnes/yr,[2] which is in general agreement with the NRCC projected estimate of 107 tonnes/yr for 1985.[10] This source is decreasing as the use of mercury in paints is regulated.

According to Voldner and Smith,[2] about 60% of the mercury consumed in the U.S. goes to landfills, leaving 40% being emitted to the atmosphere.

Fossil Fuel Combustion

Coal-Fired Power Plants

Electric utilities, including coal-fired power plants, appear to be a major source of environmental mercury. Roughly 7.0×10^7 tonnes/yr of coal are burned by electric utilities in the U.S.,[11] emitting an average of 0.18 g mercury for each tonne of coal burned,[2] or approximately 126 tonnes/yr of mercury (7.0×10^7 metric tonnes/yr of coal $\cdot$ 0.18 g Hg/tonne). This estimate is similar to Harriss and Hohenemser's estimate of 100 tonnes/yr,[12] and Voldner and Smith's[2] estimate of 113 tonnes/yr. If the total amount of mercury discharged to the environment in the U.S. from anthropogenic sources is about 790 tonnes/yr, then coal-fired power plants account for about 16% of the total anthropogenic mercury emissions to U.S. atmosphere. Based on *global* anthropogenic mercury emissions of about 3500 tonnes/yr,[13] U.S. coal-fired power plants contribute about 3.2% of total mercury emissions on a global scale, a percentage similar to Douglas' estimate of 3%.[14]

Industrial and Other Sources

Mercury emissions from coal are also attributable to industrial, residential, and commercial sectors. The industrial sector, which includes coke plants, combusts about 1.04×10^7 metric tonnes of coal/year. Assuming the same amount of mercury/tonne of coal (0.18 g Hg/tonne), the industrial sector emits 19 tonnes Hg/yr.

Other economic sectors, including transportation, residential, and commercial collectively consume about 5.4×10^5 tonnes Hg/yr. Assuming 0.18 g Hg/tonne, these sectors release about 1 tonne/yr of mercury to the atmosphere.

Petroleum Refineries

7.25×10^7 tonnes of petroleum products (crude oil, natural gas liquids, unfinished oils, and blending components) were refined in the U.S. in 1990[11] to produce motor gasoline, fuels, kerosene, and miscellaneous products. Assuming all mercury in petroleum products is released during refining or during eventual use, and assuming a concentration of mercury in oil of 0.05 mg Hg/kg,[15,16] we estimate that 36 tonnes of mercury are emitted either during refining or use of petroleum products. If the total amount of mercury discharged to the environment from anthropogenic sources is about 790 tonnes/yr in the U.S., then petroleum refineries (or subsequent use) contribute about 5% of the total emissions to the atmosphere.

Fuel Oil

1.06×10^7 tonnes of oil were combusted in the U.S. in 1989.[17] The concentration of mercury in oil is roughly 0.05 mg Hg/kg, with oil from North America having some of the lowest impurities.[15,16] It is estimated that about 5 tonnes of mercury are emitted annually from fuel oil burning (1.06×10^7 tonnes/yr of oil · 0.05 g Hg/metric tonne). If the total amount of mercury discharged to the environment from anthropogenic sources is about 790 tonnes/yr in the U.S., then fuel oil burning is not a significant source of mercury emissions to the atmosphere (0.63%), an assessment that is consistent with other studies.[13]

Smelting and Refining of Metals

An important source of mercury emissions is the smelting and refining of non-ferrous metals, such as copper, lead, and zinc. Watson[8] estimated a release of 70 tonnes/yr of mercury to all environmental media from smelting and refining, accounting for about 9% of the total anthropogenic mercury emissions in the U.S. The U.S. EPA,[18] using 1980 inventories, estimates emissions from copper, lead, and zinc smelting to be about 51 tonnes/yr, accounting for 6% of total mercury emissions.

Municipal Waste Incinerators

Many municipal waste incinerators combust solid waste containing mercury, such as batteries, electrical components, paints, and thermostats. Table 11.1 shows the amount of mercury in municipal solid waste and the percentage contribution of each category, based on 1989 estimates.[9]

Combustion of these materials results in the emission of both particulate and gaseous mercury from the stack of the incinerator,[19] with less than 20% of the emissions being trapped by pollution controls. Ironically, the more modern the incinerator and the more efficient the pollution controls (i.e., less carbon is left onto which mercury could adsorb), the less effective the incinerator is at preventing mercury releases.[15] While EPA has designated a dry scrubber followed by a fabric filter (baghouse) as the best demonstrated pollution control technology for incinerators, this determination is based on particulate control rather than mercury capture. The average capture rate for the scrubber/baghouse system is only about 50%. Many incinerators have even less efficient mercury capture because they

Table 11.1   Discards of Mercury in Products in
Municipal Solid Waste

Products	Amount (short tons)	Percent contribution
Household batteries	621.1	87.6
Electric lighting	26.7	3.8
Paint residues	18.2	2.6
Fever thermometers	16.3	2.3
Thermostats	11.2	1.6
Pigments	10.0	1.4
Dental uses	4.0	0.6
Special paper coating	1.0	0.1
Mercury light switches	0.4	0.1
Total	709.0	100.0

Table 11.2 Mercury Emissions from Municipal Waste Incinerators[a]

MWC name	Combustor type	Emissions (g/metric tonne)
Detroit	RDF[b]	0.97
West Palm Beach	RDF	0.28
Akron	RDF	0.455
Albany	RDF	2.14
Niagara	RDF	1.58
Mean	RDF	0.84
Pinellas County	MB/WW[c]	4.235
Burnaby (unit 1)	MB/WW	2.35
Dutchess County (unit 1)	MB/WW	5.4
Portland (unit 1)	MB/WW	2.75
Indianapolis	MB/WW	1.415
Long Beach	MB/WW	0.9
Mean	MB/WW	2.4
Braintree	MB[d]	0.11
Hampton	MB	10.3
Tulsa	MB	1.79
Malmo	MB	1.07
Marion County	MB	1.44
Tsushima	MB	0.45
Red Wing	MB	2.56
Mean	MB	1.2

[a] Mercury emissions from municipal waste incinerators were determined by taking the geometric mean of the totals for the different combustor types; yields 1.3 g/metric tonne.
[b] Refuse derived fuel.
[c] Mass burn/waterwall combustor.
[d] Mass burn combustor.

are fitted with electrostatic precipitators (ESPs), which remove little (10%) or no mercury.[15]

Based on EPA data,[20] White and Nebel,[21] and Idzorek,[22] the authors estimate that with current designs and pollution control equipment, incinerators emit about 1.3 g mercury for each metric tonne of municipal solid waste (MSW) combusted (Table 11.2). Roughly 180,000,000 tonnes of MSW are generated annually in the U.S., with approximately 14% by weight of this material incinerated.[23] Therefore, about 33 tonnes of mercury are emitted annually from MWCs in the U.S. (180,000,000 tonnes/yr of MSW · 0.14 · 1.3 g Hg per metric tonne). Based on this estimate, MWCs account for about 4% of the total anthropogenic mercury emissions (790 tonnes/yr) in the U.S. Collins and Cole[15] estimate MWC mercury emissions to all environmental media (air, water, land) to be about 40 tonnes/yr in the U.S., while Voldner and Smith[2] estimate that such emissions are about 68 tonnes/yr. Another type of incineration, sewage sludge incineration, is estimated by the U.S. EPA[18] to release 36 tonnes/yr. The combination of emissions from sewage sludge incineration and the authors' estimate of emissions from municipal waste incineration yields a total of 69 tonnes/yr of atmospheric mercury emissions, accounting for almost 9% of the total anthropogenic mercury emissions in the U.S.

Are Municipal Waste Incinerators a Major Source of Human Exposure to Mercury?

Municipal solid waste incineration is perceived by some to be a major source of human exposure to mercury. On a national scale, the authors estimated that MWCs contribute only 4% of mercury emissions in the U.S. As more MWCs begin operation, the amount of mercury released to the environment by this source will likely increase. Presently, however, municipal waste combustion appears to be a relatively minor source of mercury emissions in the U.S. In fact, even on a local scale, MWCs cause generally only minor, and very local, increases in atmospheric concentrations of mercury.

Atmospheric mercury concentrations dissipate quickly in the vicinity of incinerators so that ambient concentrations around incinerators are similar to background mercury concentrations. This dissipation is due in large part to precipitation events, in which most of the mercury is washed out of the atmosphere and deposited. For example, an exposure assessment study of a municipal waste combustor in Rutland, VT used an air dispersion model to predict maximum annual average concentrations of mercury in ambient air around the incinerator. The maximum predicted mercury concentration was 0.57 ng/m,³ a value below typical background concentrations (2.5 ng/m³) of mercury.[24] In addition, measured mercury concentrations in environmental media near the incinerator such as produce, forage, water, and soil were below detection limits, supporting the evidence that municipal waste incinerators are not a major source of mercury. Glass et al.[25] measured an atmospheric mercury concentration of 3.0 ng/m³ at 10 km from an operating municipal waste incinerator in Duluth, MN. This value is also within the range of background concentrations for the U.S. In a study of all U.S. incinerators that existed in 1985 and that were projected (over the next 10–15 years), Radian Corporation data[26] show that maximum ground concentrations of mercury near incinerators do not exceed the NESHAP guideline of 1 μg/m³. The highest concentration was shown to be only 40% of that guideline. The incinerators studied varied in size, location, type, and design.

Background Concentrations of Mercury

Because of increasing anthropogenic emissions of mercury, background concentrations of mercury are increasing. Mercury concentrations in precipitation, air, vegetation, water, soil, particulates, and deposition have been documented in the literature. These data show that background concentrations of mercury in environmental media are an important source of human exposure to mercury.

Mercury Concentrations in Fish

Mercury contaminates fish in hundreds of lakes and rivers in the upper Midwest, around the Great Lakes, and in other states and provinces.[27] In remote Michigan lakes, 15% of fish contain mercury in excess of the state health advisory level of 0.5 ppm.[28] Approximately 30% of Wisconsin lakes and 50% of Florida lakes analyzed (about 100 major systems) contain fish with mercury levels exceeding state health standards.[29] Currently, 20 states have issued fish advisories for mercury. Background mercury concentrations (wet weight) in fish range from about 20 to about 1500 ppb.[30-32] We select a median concentration of mercury in fish of 350 ppb (wet weight) as representative of the U.S.

The principal sources of mercury input to rural lakes are overland transport from watersheds and direct atmospheric deposition to lake surfaces. Both sources are important and together can account for mercury levels found in lakes.[33] Swain et al.[34] report that about 20% of mercury in lakes is attributable to overland runoff and about 80% is attributable to direct mercury deposition. The ratio of runoff loading to direct deposition ranges from about 0.6 for clearwater lakes to about 6.0 for brownwater lakes (lakes high in organic matter). This ratio changes during the year with changing environmental conditions, with direct deposition dominating under dry conditions.[35] Mercury in runoff water is found primarily attached to soil organic matter, while in precipitation, mercuric oxide dominates. In contrast to other metals such as cadmium and zinc, acidification of soils and waters does not appear to increase mercury transport from watershed to lake.[35] However, mercury's affinity for organic matter in soil may increase as precipitation pH decreases.[34]

Are Background Levels of Mercury Increasing?

Historical records in soils, peats, ice sheets, sediments, and tree rings in most parts of the Northern Hemisphere show that metal pollution has been increasing since the beginning of the Industrial Revolution; mercury is no exception. Lake sediment cores show that atmospheric mercury deposition has increased a factor of three or four since precolonial times.[34,36,37] Significant increases of mercury in lake sediment cores have also been found in Canada[38] and the U.S.[39,40] Lindqvist[33] estimates that sediment concentrations in lakes are about five times higher than background levels and that concentrations in fish have increased by a factor of four to six. Slemr and Langer[41] estimate that atmospheric concentrations of mercury have increased by 1.46% per year in the Northern Hemisphere over the past 20 years. The increase in atmospheric mercury concentrations and in mercury deposition in remote areas[34] implies that regional and global emissions of mercury are being deposited on remote sites.

The observed four- to fivefold increase in environmental levels of mercury has a direct bearing on the question of the relative contribution of natural versus anthropogenic sources to ambient levels of mercury. Until recently, it was the prevailing opinion that natural sources of mercury dominated anthropogenic sources by a factor of from two to four.[16] However, if total environmental releases of mercury have increased by a factor of four, then anthropogenic sources must now contribute four times more than natural sources, or about 80% of the total. This is consistent with Slemr and Langer's[41] estimate of 75%.

Human Exposure to Mercury

Human exposure to mercury occurs primarily via indirect pathways such as food ingestion, rather than direct pathways (e.g., inhalation). The authors used a terrestrial food chain model to predict mercury concentrations, based on ambient mercury concentration data. Since mercury concentrations near incinerators are close to ambient levels, the results from this analysis could be comparable to human exposure estimates for populations near incinerators. Results indicate that indirect exposures contribute about 99% of the total daily intake of mercury, while direct pathways contribute only about 0.7%. Mercury intake via fish ingestion is the major route of exposure (58%). Surprisingly, vegetation ingestion is also a major contributor to daily intake of mercury (31%). Total daily intake of mercury was estimated to be 11 μg/d, which agrees well with previous estimates ranging from 5–20 μg/d.[3,10,42]

DIOXIN

Sources of Dioxin

While TCDD emissions from various sources have been estimated or measured, the relative contribution of these sources to the total TCDD load is unknown. By using a Fugacity Food Chain (FFC) model[43,44] in a preliminary attempt to determine the rate of environmental input of TCDD necessary to maintain measured background levels of TCDD in air and soil, Travis and Hattemer-Frey[45] estimated TCDD emissions into the contiguous U.S. to be 80 kg/yr, with a range of 25 to 120 kg/yr.[46] The major sources of dioxin discussed in the following sections have readily available emissions data, and approximations are limited by the accuracy of reported data. In addition to the sources discussed below, other potential sources of TCDD include: forest fires; discharges from metal processing and treatment plants and copper smelting plants; chemical/industrial incineration; pentachlorophenol production; and waste manufacture, use, and disposal.

In this paper, sources and human exposure were investigated only for the TCDD compound because it is the most toxic of the dioxins and furans. However, other dioxins and furans also contribute to the loading of total dioxins into the environment.

Municipal Solid Waste Incinerators

Travis and Hattemer-Frey[45] found TCDD emissions from existing U.S. MSW incinerators (at 12% CO_2) to be lognormally distributed with a geometric mean of 2.4 × 10^{-8} g/s. Assuming 130 operating MSW incinerators in the U.S.,[47] total emissions from MSW incinerators are approximately 0.1 kg/yr, or less than 1% of total TCDD environmental input. Assuming an empirical Toxic Equivalent (TEQ) to 2,3,7,8–TCDD emissions ratio of 7 to 1, the total TEQ emissions from MSW incinerators in the U.S. is approximately 600 g/yr. This compares with recent estimates of 570 and 380 g/yr.[48] Assuming only 380 g/yr of dioxin are emitted, the equivalent amount of TCDD is 0.05 kg/yr.

Motor Vehicles

Direct and indirect evidence indicates that motor vehicles are a prominent source of environmental TCDD.[49-51] Although direct measurements of TCDD emissions from motor vehicles are scarce, Marklund et al.[50] reported that cars run on leaded gasoline in Sweden emitted <2.0–12 ng TCDD/mile driven (1 mile = 1.61 km). TCDD emissions from cars run on unleaded gasoline were nondetectable. Assuming 42 million cars in the U.S. in 1985 used leaded gasoline and that the typical car was driven 7,900 miles/year,[52] Travis and Hattemer-Frey[45] estimate emissions of TCDD from vehicles run on leaded gasoline to range from <0.7 to 4.0 kg/yr, which would account for up to 5% of total TCDD input.

Hospital Waste Incinerators

Data on polychlorinated dibenzo-p-dioxins and furans (PCDD/PCDF) emissions from hospital waste incinerators are extremely limited. The California Air Resources Board (CARB) tested pollutant emissions from hospital waste incinerators in California. TCDD emissions from two incinerators in California ranged from <0.02 to 0.5 and from <0.11 to 0.24 ng/s.[53,54] Assuming that 6,185 U.S. hospitals have an incinerator on site[55] and that all incinerators are emitting TCDD at a similar rate, Travis and Hattemer-Frey[45] estimate total TCDD emissions from hospital incinerators in the U.S. appear to account for <1% of total TCDD input to the environment.

Residential Wood Burning

Residential wood combustion (RWC) has been shown to be a significant source of environmental pollution.[56] Nestrick and Lamparski[57] surveyed residential wood combustion units and found TCDD emissions ranged from nondetectable (<0.07 ng/kg) to 20 ng TCDD/kg wood burned. Assuming that in 1980–81, U.S. households burned 42 million cords of wood and that 0.677 cords equals 1 tonne,[58] Travis and Hattemer-Frey[45] estimate total TCDD emissions from RWC units to be approximately <0.06 to 1.1 kg/yr or about 1% of total TCDD input to the environment.

Pulp and Paper Mill Effluent

Pulp and paper mill liquid effluents were tested at five bleached kraft mills in the U.S. as part of the EPA/Paper Industry Dioxin Screening Survey.[59] Mean TCDD levels ranged from 0.02 to 1.1 ng/L. Assuming that 1×10^{12} gallons of effluent are released annually from U.S. mills[60] and that all effluents are contaminated at the mean level (0.15 ng/L), Travis and Hattemer-Frey[45] estimate TCDD emissions from pulp and paper mills in the U.S. to be about 0.6 kg/yr, which would account for <1% of total TCDD input into the environment.

Are Municipal Waste Incinerators the Major Source of Human Exposure to Dioxin?

Although MSW incineration has been perceived to be a major source of human exposure to TCDD, Travis and Hattemer-Frey[45] have shown that emissions from MSW incinerators on the national level account for less than 1% of total current TCDD input into the environment.

On the local level, human exposure to dioxins emitted from MSW incineration is small relative to exposure to background environmental contamination.[61] The incremental air concentration predicted to occur at a point of maximum individual exposure near a typical, modern MSW incinerator is 11 fg TEQs/m³,[62,63] about eight times lower than mean background levels of PCDDs and PCDFs measured in urban air (8 fg/m³). Edgerton et al.[64] measured background levels and concentrations of PCDDs and PCDFs in air 2 km downwind from an MSW incinerator in an urban area of Ohio. Their data show that levels near the incinerator (106 fg/m³) are virtually identical to background concentrations (103 fg/m³). Thus, PCDD/PCDF concentrations around operating MSW incinerators are not substantially elevated relative to mean background urban air levels.

Hattemer-Frey and Travis[65] predicted the maximally exposed individual's (MEI) daily intake of PCDDs and PCDFs living near a typical, modern MSW incinerator by taking the geometric mean of the total daily intake estimates reported in risk assessments prepared for 12 proposed incinerators from all pathways. The predicted daily intake of PCDDs and PCDFs (expressed in TEQs) by the MEI is 130 times *less than* exposure to background contamination.

These data show that human exposure to dioxin from an incinerator is not excessive relative to exposure to background levels. In addition, the individual lifetime cancer risk associated with exposure to facilty-emitted dioxin, 2×10^{-6}, is two orders of magnitude lower than the cancer risk associated with exposure to background environmental TCDD contamination (2×10^{-4}).

Rappe et al.[70] found that the isomeric pattern of dioxins and furans in environmental media is virtually identical to the pattern established for several incineration sources, including MSW incinerators, hazardous waste incinerators, steel mills, copper smelting plants, and motor vehicle exhausts. Environmental concentrations of dioxin cannot, therefore, be linked to any one combustion source.

Background Concentrations of Dioxin

Background concentrations of TCDD in environmental media are an important source of human exposure to dioxin. Measured atmospheric concentrations range from 17 to 305 fg/m^3, with a geometric mean of 86 fg/m^3. TCDD levels in soil from Great Britain (9.4 ng/kg),[71] West Germany (2.4 ng/kg),[70] and the U.S. (1.0 ng/kg)[72] vary by less than one order of magnitude. Nielsen and Lokke[73] measured soil samples from three sites 0.1–0.5 km from a small MSW incinerator and concluded that PCDD/PCDF concentrations near that facility were not elevated relative to background concentrations.

Human Exposure to Dioxin

Human exposure to dioxin occurs primarily via indirect pathways such as food ingestion, rather than direct pathways. We used a fugacity food chain model[43,44] to predict TCDD concentrations in contaminated agricultural products. Results indicate that the food chain, especially meat and dairy products, contribute about 99% of the total daily intake of TCDD, while inhalation and ingestion of water, soil, eggs, and produce are not major exposure pathways. Total daily intake was estimated to be 34.8 pg/d.

CONCLUSIONS

Examination of the environmental partitioning of mercury and dioxin has yielded several significant findings. The greatest contributors to annual mercury emissions in the U.S. are commercial and consumer uses such as the chemical industry, paints, and electrical industries (61%), followed by coal-fired power plants (16%), and smelting and refining (6%). Municipal waste incineration is not a major source, contributing only 4%. Anthropogenic sources of mercury have increased background levels four- to fivefold since colonial times, implying that anthropogenic sources now contribute 80% of total environmental mercury pollution.

It has recently been noticed that fish mercury concentrations in hundreds of remote U.S. lakes are elevated. Twenty states have consequently issued fish advisories for mercury. The principal source of mercury input to these rural lakes is direct atmospheric deposition onto the lakes and surrounding watersheds. Soil organic matter in the surrounding watersheds acts as a carrier of mercury during overland transport. Mercury levels in fish are largely caused by atmospheric transport and deposition of mercury into lakes. For this reason, atmospheric emissions of mercury by all sources, especially coal-fired power plants and municipal waste combustors should be more effectively controlled.

Ambient measurements confirm that environmental TCDD contamination is widespread. MSW incinerators, automobiles, residential wood combustion, hospital waste incinerators, and pulp and paper mills together account for a maximum of 6% of current TCDD input into the U.S. environment. While there are limitations inherent in extrapolating data from a small sample size, a more focused effort is needed to identify the major source(s) of TCDD input so that background levels and human exposure can be minimized.

The fact that the five TCDD sources characterized here and elsewhere[45] together account for a maximum of 11% of total annual TCDD input into the U.S. environment may indicate various things: (1) some unidentified major sources of TCDD contamination exist, or (2) multiple environmental sources of TCDD exist and no one source dominates total input to the environment.

Further research is needed to more accurately determine the relative contribution of source to the total TCDD environmental input.

REFERENCES

1. J.O. Nriagu, Global Metal Pollution: Poisoning the Biosphere? *Environment.* 32(7):7–33 (1990).
2. E. Voldner and L. Smith, Production, Usage, and Atmospheric Emissions of 14 Priority Toxic Chemicals, in Appendix 2 of *Proceedings of the Workshop on the Great Lakes Atmospheric Deposition*, International Air Quality Advisory Board of the International Joint Commission, October 29–31, 1986.
3. S. Mitra, *Mercury in the Ecosystem: Its Dispersion and Pollution Today*, TransTech Publications, Switzerland, 1986.
4. U.S. EPA, *Health Assessment Document for Polychlorinated dibenzo-p-dioxins.* EPA/600/8-84-014F. Washington, D.C., 1985.
5. C.C. Travis, H.A. Hattemer-Frey, and E. Silbergeld, Dioxin, Dioxin Everywhere. *Environ. Sci. Technol.* 23: 1061–1063, 1989.
6. B.D. Eitzer and R.A. Hites, Dioxins and Furans in the Ambient Atmosphere: A Baseline Study. *Chemosphere* 18:593–598, 1989.
7. U.S. EPA, *Methodology for Assessing Health Risks Associated with Indirect Exposure to Combustor Emissions*, EPA 600/6-90/003, Office of Health and Environmental Assessment, Washington, D.C., 1990.
8. W.D. Watson, In *The Biogeochemistry of Mercury in the Environment*, J.O. Nriagu, ed. Elsevier-North Holland, Amsterdam.
9. U.S. EPA, *Characterization of Products Containing Mercury in Municipal Solid Waste in the U.S., 1970 to 2000*, PB92-162569, 1992.
10. NRCC (National Research Council of Canada), *An Assessment of Mercury in the Environment*, National Academy of Science Printing and Publishing Office, Washington, D.C., 185 pp. 1978.
11. Annual Energy Review 1990, Energy Information Administration, Washington, D.C. DOE/EIA-0384(90), May 1991.
12. R.C. Harriss and C. Hohenemser, Mercury: Measuring and Managing the Risk, *Environment*, 20: 25–26, 1978.
13. J.O. Nriagu and J.M. Pacyna, Quantitative Assessment of Worldwide Contamination of Air, Water, and Soil by Trace Metals, *Nature,* 333:134–139, 1988.

14. J. Douglas, Mercury in the Environment, *EPRI Journal*, 16(8):4–11, 1991.
15. R. Collins and H.S. Cole, *Mercury Rising*, Clean Water Action, Clean Water Fund, Research and Technical Center, Washington, D.C., 1990.
16. T.C. Hutchinson and K.S. Meema, Ed., *Lead, Mercury, Cadmium, and Arsenic in the Environment*, John Wiley and Sons, New York, 1987.
17. *Energy Statistics Yearbook 1989*, Department of International Economic and Social Affairs, United Nations, New York, 1991.
18. U.S. EPA, *An Exposure and Risk Assessment for Mercury*, EPA-440/4-85-011, 1981.
19. Y. Otani, C. Kanaoka, C. Usui, S. Matsui, and H. Emi, Adsorption of Mercury Vapor on Particles, *Env. Sci. Technol.*, 20(7): 735–738, 1986.
20. U.S. EPA, *Municipal Waste Combustion Study: Emission Data Base for Municipal Waste Combustors*, EPA/530-SW-87-0216, 1987.
21. D.M. White and K.L. Nebel, *Summary of Information Related to Mercury Emission Rates and Control Technologies Applied to Municipal Waste Combustors*, Radian Corporation, Research Triangle Park, N.C., 1990.
22. J.F. Idzorek, *Municipal Solid Waste Incineration: Emission Testing Experiences in Minnesota*, Minnesota Pollution Control Agency, Air Quality Division, 1990.
23. *Journal of the Air and Waste Management Association*, 41(3):259–60, 1991.
24. C. Sonich-Mullin and R.J.F. Bruins, *Feasibility of Environmental Monitoring and Exposure Assessment for a Municipal Waste Combustor: Rutland, Vermont Pilot Study*. ECAO, U.S. EPA, 1991.
25. G.E. Glass, J.A. Sorenson, K. W. Schmidt, and G.R. Rapp, Jr., New Source Identification of Mercury Contamination in the Great Lakes, *Env. Sci. Technol.*, 24(7): 1059–1069, 1990.
26. Radian Corporation, *Municipal Waste Combustion Study: Assessment of the Health Risks Associated with Municipal Waste Combustion Emissions*, for the Office of Air Quality Planning and Standards, EPA, 1989.
27. J.A. Sorenson, G.E. Glass, K.W. Schmidt, J.K. Huber, and G.R. Rapp, Jr., Airborne Mercury Deposition and Watershed Characteristics in Relation to Mercury Concentrations in Water, Sediments, Plankton, and Fish of Eighty Northern Minnesota Lakes, *Env. Sci. Technol.*, 24(11): 1716–1727, 1990.
28. D. Porcella, *EPRI Journal* April/May: 46–49, 1990.
29. Wisconsin Department of Natural Resources, Health Advisory for People Who Eat Sport Fish From Wisconsin Water, Wisconsin Division of Health: Madison, WI, 1987.
30. D.C. Adriano, *Trace Elements in the Terrestrial Environment*, Springer-Verlag, New York, 1986.
31. T.M. Grieb, C.T. Driscoll, S.P. Gloss, C.L. Schofield, G.L. Bowie, and D.B. Porcella, Factors Affecting Mercury Accumulation in Fish in the Upper Michigan Peninsula, *Env. Toxicol. Chem.* 9: 919–930, 1990.
32. Van Horn, *Materials Balance and Technology Assessment of Mercury and Its Compounds on a National and Regional Basis*, EPA-560/3-75-007, 1975.
33. O. Lindqvist, Ed., Mercury in the Swedish Environment: Recent Research on Causes, Consequences, and Corrective Methods, In *Water, Air, Soil Poll.* 55(1, 2), 1991.
34. E.B. Swain, D.R. Engstrom, M.E. Brigham, T.A. Henning, and P.L. Brezonik, Increasing Rates of Atmospheric Mercury Deposition in Midcontinental North America, *Science* 257:784–787, 1992.

35. S.C. Barton, N.D. Johnson, and J. Christison, Atmospheric Mercury Deposition in Ontario, *Proceedings of the Air Pollution Control Association, 74th Annual Meeting*, Paper JAPCA 81-60.4, Philadelphia, June 21–26, 1981.

36. R.G. Rada, J.G. Wiener, M.R. Winfrey, and D.E. Powell, Recent Increases in Atmospheric Deposition of Mercury in North-Central Wisconsin Lakes Inferred from Sediment Analyses, *Arch. Env. Contam. Toxicol.* 18: 175–181, 1989.

37. R. Douglas Evans, *Arch. Env. Contam. Toxicol.* 15: 505–512, 1986.

38. M. Ouellet and H.G. Jones, Historical Changes in Acid Precipitation and Heavy Metals Deposition Originating from Fossil Fuel Combustion in Eastern North America as Revealed by Lake Sediment Geochemistry, *Wat. Sci. Technol.* 15: 115–130, 1983.

39. J.J. Akielaszek and T.A. Hanes, Mercury in the Muscle Tissue of Fish from Three Northern Maine Lakes, *Sci. Tot. Env.* 27: 201–208, 1981.

40. M. Heit, Y. Tan, and C. Klusek, Anthropogenic Trace Elements and Polycyclic Aromatic Hydrocarbon Levels in Sediment Cores from Two Lakes in the Adirondack and Acid Lake Region, *Water, Air, Soil Poll.* 15: 144–464, 1981.

41. F. Slemr and E. Langer, Increase in Global Atmospheric Concentrations of Mercury Inferred from Measurements over the Atlantic Ocean, *Nature* 355:434–437, 1992.

42. L.T. Friberg and J. Vostal, *Mercury in the Environment*. CRC Press, Cleveland, Ohio, 1972.

43. C.C. Travis and H.A. Hattemer-Frey, Human Exposure to 2,3,7,8–TCDD, *Chemosphere* 16: 2331–1342, 1987.

44. H.A. Hattemer-Frey and C.C. Travis, Pentachlorophenol: Environmental Partitioning and Human Exposure, *Arch. Environ. Contam. Toxicol.* 18: 482–489, 1989.

45. C.C. Travis and H.A. Hattemer-Frey, Human Exposure to Dioxin, *Sci. of the Total Environ.* 104: 97–127, 1991.

46. C.C. Travis and H.A. Hattemer-Frey, TCDD Contamination of Fish and the Potential for Human Exposure. *Environ. Int.* 16: 155–162, 1990.

47. IRR (Institute of Resource Recovery). *Resource Recovery Focus*, 1: 3, 1986.

48. K.H. Jones, Diesel Truck Emissions: An Unrecognized Source of PCDD/PCDF Exposure in the U.S. *Journal of Risk Analysis*. In press.

49. K.H. Ballschmiter, R. Buchert, R. Niemczyk, A. Munder, and M. Swerev, Automobile Exhausts versus Municipal Waste Incineration as Sources of the Polychloro-dibenzodioxins (PCDD) and -furans (PCDF) Found in the Environment, *Chemosphere* 15: 901–915, 1986.

50. S. Marklund, C. Rappe, and M. Tysklind. Identification of Polychlorinated Dibenzofurans and Dioxins in Exhausts from Cars Run on Leaded Gasoline, *Chemosphere* 16: 29–36, 1987.

51. C. Rappe, L.-O. Kjeller, P. Bruckmann, and K.-H. Hackhe. Identification and Quantification of PCDDs and PCDFs in Urban Air. *Chemosphere* 17: 3–20, 1988.

52. USDC (U.S. Department of Commerce). Statistical Abstract of the U.S., U.S. Government Printing Office, Washington, D.C., 1988.

53. A. Jenkins, C. Castronovo, G. Linder, P. Ouchida, and D.C. Simeroth, Evaluation Test on a Hospital Refuse Incinerator at Cedars Sinai Medical Center, Los Angeles, Report No. ARB/SS-87-11. California Air Resources Board, Sacramento, CA. 1987.

54. A. Jenkins, P. Ouchida, and G. Lew, Evaluation Retest on a Hospital Refuse Incinerator at Sutter General Hospital, Sacramento, CA. Report No. ARM/ML-88-026. California Air Resources Board, Sacramento, CA, 1988.

55. U.S. EPA, Hospital Waste Combustion Study Data Gathering Phase Final Report. EPA-450/3-88-017. Office of Air Quality Planning and Standards, Research Triangle Park, NC. 1988.

56. J.A. Cooper, Environmental Impact of Residential Wood Combustion Emissions and Its Implications, *J. Air Pollut. Control Assoc.* 30: 855–861, 1980.

57. T.J. Nestrick and L.L. Lamparski, Isomer-Specific Determination of Chlorinated Dioxins for Assessment of Formation and Potential Environmental Emission from Wood Combustion. *Anal. Chem.* 54: 2292–2299, 1982.

58. K.E. Skog and I.A. Watterson, Residential Fuelwood Use in the U.S.: 1980–81. Report No. ADA-131724. U.S. Department of Agriculture. Forest Products Laboratory, Madison, WI, 1983.

59. G. Amendola, D. Barna, R. Blosser, L. LaFluer, A. McBride, F. Thomas, T. Tiernan, and R. Whittmore, The Occurrence and Fate of PCDDs and PCDFs in Five Bleached Kraft Pulp and Paper Mills, *Chemosphere* 18: 1181–1188, 1989.

60. I. Gelman, *PCDDs and PCDFs Sources and Releases to the Environment,* National Council of the Paper Industry for Air and Stream Improvement, Inc. New York, NY., 1989.

61. C.C. Travis and H.A. Hattemer-Frey, A Perspective on Dioxin Emissions from Municipal Solid Waste Incinerators, *Risk Anal.* 9: 91–97, 1989.

62. C.C. Travis and H.A. Hattemer-Frey, Human Exposure to Dioxin from Municipal Solid Waste Incineration, *Waste Mgmt.* 9: 151–156, 1989.

63. H.A. Hattemer-Frey and C.C. Travis, A Perspective on Municipal Solid Waste Incinerators as a Source of Environmental Dioxin, In C.C. Travis and H.A. Hattemer-Frey (eds.), *Municipal Waste Incineration and Human Health,* CRC Press, Boca Raton, FL., 1990.

64. S.A. Edgerton, J.M. Czuczwa, J.D. Rench, R.F. Hodanbosi, and P.J. Koval, Ambient Air Concentrations of Polychlorinated Dibenzo-p-dioxins and Dibenzofurans in Ohio: Sources and Health Risk Assessment, *Chemosphere* 18: 1713–1730, 1989.

65. H.A. Hattemer-Frey and C.C. Travis, Characterizing the Extent of Human Exposure to PCDDs and PCDFs Emitted from Municipal Solid Waste Incinerators. In: C.C. Travis and H.A. Hattemer-Frey (eds.), *Municipal Waste Incinerators and Human Health,* CRC Press, Boca Raton, FL, 1990.

66. H. Beck, W. Eckart, W. Mathar, and R. Wittowski, PCDD and PCDF Body Burden from Food Intake in the Federal Republic of Germany, *Chemosphere* 18: 587–592, 1989.

67. M. Ono, Y. Kashima, T. Wakimoto, and R. Tatsukawa, Daily Intake of PCDDs and PCDFs by Japanese Through Food, *Chemosphere* 16: 1823–1828, 1987.

68. OME (Ontario Ministry of the Environment). Polychlorinated Dibenzo-p-dioxins and Polychlorinated Dibenzofurans and Other Organochlroine Contaminants in Food, Ministry of Agriculture and Food and Environment, Toronto, Ontario, Canada, 1988.

69. P. Furst, C. Furst, and W. Groebel, Levels of PCDDs and PCDFs in Foodstuffs from the Federal Republic of Germany, *Chemosphere* 20: 787–792, 1990.

70. C. Rappe, R. Andersson, P.-A. Bergqvist, C. Brohede, M. Hansson, L.-O. Kjeller, G. Lindstrom, S. Marklund, M. Nygren, S.E. Swanson, M. Tyskline, and K. Wiberg, Overview on Environmental Fate of Chlorinated Dioxins and Dibenzofurans: Sources, Levels and Isomeric Pattern in Various Matrices, *Chemosphere* 16: 1603–1618, 1987.

71. C.S. Creaser, A.R. Fernandes, A. Al-Haddad, S.J. Harrard, R.B. Homer, P.W. Skett, and E.A. Cox, Survey of Background Levels of PCDDs and PCDFs in UK Soils, *Chemosphere* 18: 767–776, 1989.

72. T.J. Nestrick, L.L. Lamparski, N.M. Frawley, R.A. Hummel, C.W. Kocher, N.H. Mahle, J.W. McCoy, D.L. Miller, T.L. Peters, J.L. Pillepich, W.E. Smith, and S.W. Tobey, Perspectives of a Large Scale Environmental Survey for Chlorinated Dioxins: Overview and Soil Data, *Chemosphere* 15: 1453–1460, 1986.

73. P.G. Nielsen and H. Lokke, PCDDs, PCDFs, and Metals in Selected Danish Soils, *Ecotoxicol. Environ. Saf.* 14: 147–156, 1987.

CHAPTER 12

Municipal Solid Waste Combustion Ash: State-of-the-Knowledge

Carlton C. Wiles

INTRODUCTION

Over the past several years there has been significant controversy concerning the proper management of the residues from combusting municipal solid waste (MSW) and their regulatory classification as hazardous or non-hazardous waste. This controversy and other factors (e.g., lack of federal guidance, heavy metal content, etc.) have resulted in inconsistent management requirements among several states and uncertainty about beneficial utilization of the residues. Heavy metal content and leaching of these metals (especially in the TCLP test) is most often cited as the reasons the material should be managed as a hazardous waste. If not managed properly, contamination of groundwater by leaching of soluble salts from the ashes may also be a concern. The U.S. lags behind some countries in ash utilization. Although research and demonstration projects have indicated that the ashes can be beneficially utilized, less than 5% of the ashes are utilized in the U.S. Other countries, including Denmark, the Netherlands, France, Germany, Switzerland, and Japan are further advanced in ash utilization and in establishment of a systematic process for evaluating and selecting disposal and utilization options. This chapter discusses ash characteristics, the state of ash management in the U.S., federal initiatives, results of laboratory and field characterization of leachates from the ashes, barriers to ash utilization in the U.S., and international perspectives.

In 1993 there were 125 waste-to-energy (WTE) facilities operating in the U.S. The combined design capacity of these plants ranges from 99,400 to over 107,000 tons per year (tpy), with an estimated electrical generating capacity of 1,800 to more than 2,900 megawatts (MW).[1,2] The combined estimated quantities of ash produced by these facilities is approximately 8.5 million to 9 million tons/year.

1-56670-215-1/97/$0.00+$.50
© 1997 by CRC Press, Inc.

This will increase to as much as 17,000,000 tons a year or more in the future as more WTE facilities are placed into service. This ash must be managed on a daily basis.

Municipal Waste Combustion (MWC) residues are generated at several points in the process of burning MSW for energy recovery. Solids retained on furnace grates following combustion and solids passing through the grates (siftings) are generally referred to as bottom ash. Entrained particulates that are trapped and residues generated by acid gas scrubbers and subsequently removed by fabric filters and/or electrostatic precipitators (ESPs) are normally referred to as air pollution control (APC) residue. In some cases, especially in Europe, ESPs are used to remove particulates before wet scrubbers. This stream may be considered as an APC residue or as a fly ash. Entrained particulates and condensed vaporized metals trapped in heat exchangers generate a small quantity of ash, referred to as heat recovery ash. Heat recovery ash is combined with either the APC residue or the bottom ash. Approximately 80% of the residues generated are bottom ash. In the U.S., these fractions are normally collected together as a combined ash.

The physical characteristics of bottom ash resembles an aggregate while the APC residue is much finer. The major elements in bottom ash are aluminum, calcium, iron, silicon, and chlorine. Major elements in the APC residues are aluminum, calcium, potassium, silicon, sodium, zinc, and chlorine. Most metals are present as oxides. The APC residue contains significantly higher concentrations of cadmium, lead, and zinc than does bottom ash. The APC fraction also contains higher concentrations of soluble salts.

Leaching of cadmium and lead from the residues in laboratory tests, particularly the Toxicity Characteristics Leaching Procedure (TCLP),[3] has been the key issue regarding the classification and management of these materials. Concentrations of heavy metals in leachates from laboratory tests are normally much higher than the metal concentrations in the leachates from ash monofills. There is disagreement among the technical community, regulators, environmental groups, and others concerning the use of laboratory tests to predict the ultimate fate of metals when the ashes are placed into the environment, either for utilization or disposal. Rather than a single test, one can learn much more from several tests designed to determine concentrations in the ashes, the amounts available for release under worst case conditions, and the amount expected to be released over time under the actual conditions of utilization and disposal. Only a small fraction of the metals present in the ashes are normally leached. Soluble salts, on the other hand, are almost all released and should be properly managed.[4]

Most MWC residues generated in the U.S. are disposed into monofills lined with either clay soil liners or synthetic liners or both. The designs usually include provisions for leak detection and leachate collection. While some other countries utilize a significant portion of the residues, probably less than 5% are used in the U.S. The debate over classification and concern about the release of heavy metals combined with a lack of federal guidance and consideration of related issues have impeded utilization in the U.S.[4-7]

There are options for using ash residues and for treating them prior to use or as a requirement for disposal. Treatment options include processing to remove ferrous metals, compaction, aging during storage, solidification/stabilization, vitrification, and chemical extraction. Major utilization options include aggregate for road base, embankments, asphalt pavements, and aggregate in Portland cement for construction. Research and field projects to demonstrate these various uses have been conducted or are underway or are planned.[5-9] In order for utilization to proceed, results from these projects must demonstrate no adverse effect on the environment or human health. Additionally, criteria and technical guidance to assure safe use must be developed.

DEFINITION OF MSW COMBUSTION ASHES AND RESIDUES

The ashes from combustion of MSW are routinely classified into three categories: bottom ash, APC residues, and combined (i.e., the combination of bottom ash and APC residues). In the U.S. these streams are normally combined for disposal, while in most European countries and in Canada they are separated into APC residues and bottom ash fractions.[10]

The terms ash and residue are often used interchangeably in this chapter; however, interchangeable use can be misleading. Ash refers to material remaining after complete combustion of materials while residue refers to unburnt material, scrubber sludge, reaction products and similar materials that end up in the ash/residue streams. Figure 12.1 depicts points in a combustion facility from which ash/residue streams are generated.[10] Table 12.1 provides a technically correct description of the various fractions of the ashes and residue. Bottom ash comprises

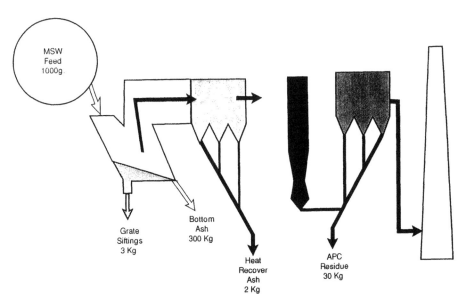

Figure 12.1 Sources of residues in the MSW combustion process. (Adapted from the International Ash Working Group.)

Table 12.1   Description of Ash Fractions Generated from the Combustion of MSW

Ash fraction	Description	Comments
Bottom	Material discharged from the bottom of the furnace, primarily the grate.	Normally the term bottom ash also includes grate siftings.
Grate siftings or riddlings	Material falling through the furnace grates.	Generally combined with bottom ash in the quench system.
Heat recovery ash (HRA)	Particulate matter collected from the heat recovery system.	May be further subdivided into boiler ash, economizer ash, or superheater ash, etc. depending on the area of the heat recovery system from which it was collected.
Fly ash	Particulate matter carried over from the furnace and removed from the flue gas prior to injection of sorbents.	Includes volatiles condensed during flue gas cooling. Excludes ashes from the heat recovery system.
APC residue	Combined material collected in the air pollution control devices, including fly ash, injected sorbents, and flue gas condensate.	
Combined ash	Mixture of bottom ash, grate siftings, and APC residues.	Waste-to-energy facilities in the U.S. routinely manage combined ash. Bottom ash and APC residues are collected and managed separately in Canada and Europe.

the major portion of the residues generated, depending on the combustion facility design, operating conditions, and characteristics of the waste being combusted.

Physical Characteristics

Bottom Ash — Bottom ash is a heterogeneous mixture of slag, ferrous and non-ferrous metals, ceramics, glass, other non-combustibles, and uncombusted organics. Up to 20% of the bottom ash has a particle size of >10 cm, consisting of ferrous and non-ferrous metals, slags, and construction-type materials. The fraction <10 cm is somewhat uniform with up to 10% fines <200 mm. This latter fraction resembles well graded sand and gravel. Bottom ash is a very porous lightweight aggregate with high specific surface areas. It has dry densities of 950 kg/cm^3 or higher, with specific gravities of 1.5 to 2 for the fine fraction and 1.8 to 2.4 for the coarse fraction.

Bottom ash contains varying amounts of moisture as a result of quenching. In modern WTE facilities, the geotechnical water content normally ranges from 15 to 25% (wet weight/dry weight). Higher levels may be found in older systems, and lower levels in mass burn facilities equipped with ram dischargers. This moisture content is important for fugitive dust control and as an aid in compaction. With optimum moisture content of approximately 16%, the bottom ash can be compacted to Proctor densities of 1600 kg/m^3 resulting in hydraulic conductivities of 10^{-6} cm/sec or less. Such hydraulic conductivities can play an important role in the management and utilization of these residues.

APC Residues — The residues from dry/semi-dry APC systems are fine particulate mixtures of fly ash consisting of reaction products of primarily calcium chlorides and unreacted lime used for acid gas emission controls. The fly ash has the largest particle size, followed by the residues from the electrostatic precipitators and the fabric filters. APC residues are usually highly soluble in water (25 to 85% by weight) due to their high concentrations of soluble salts.

Chemical Characteristics

The chemical characteristics of ashes and residues are the major reason for the concern regarding their classification as hazardous or non-hazardous wastes and their ultimate management requirements. Ash and residue fractions contain varying amounts of trace metals and soluble salts which potentially could result in adverse environmental effects if improperly managed. Cadmium and lead have represented major areas of concern; more recently some concern has also been expressed regarding the aquatic toxicity of copper and mercury, due to more stringent controls on mercury emissions. Table 12.2 provides the ranges of constituents found in bottom ash, fly ash, and wet and dry APC system residues from many facilities.[10]

Approximately 80 to 90% by weight of the bottom ash consists of O, Si, Fe, Ca, Al, Na, K, and C. Minor elements (1000 mg/kg to 10,000 mg/kg) are Mg, Ti, Cl, Mn, Ba, Zn, Cu, Pb and Cr. Trace elements (i.e., <1000 mg/kg) consist of Sn, Sb, V, Mo, As, Se, Sr, Ni, Co, Ce, Ag, Hg, B, Br, F, and I. The composition of major and most minor elements is similar to basaltic and other geologic materials. Some of the minor elements and many of the trace elements (e.g., Pb, Cu, Fn, Cd, and Hg) are enriched in the bottom ash. Grate siftings comprise approximately 1 to 3% by weight of the bottom ash. This fraction, however, contributes a significant fraction of the elemental Pb, Al, Cu, and Zn to the bottom ash of mass burn systems. The ranges of these elements in the grate siftings of mass burn systems has been reported as Pb (–5,600–34,000 mg/kg), Al (38,000–63,000 mg/kg), Cu (2,400–25,000 mg/kg), and Zn (2,450–5,300 mg/kg). These metals are largely present in their elemental form and may enter into redox reactions, generating H_2 and causing swelling. Investigations have indicated that the grate siftings contain almost 50% of the lead in the residues considered available for leaching.[10]

Total dissolved solids content (e.g., from the dissolution of NaCl, $CaCl_2$, $CaSO_4$) is fairly low in bottom ash, ranging from 3–14 wt % in modern mass burn facilities.

The major elemental constituents in APC residues are O, Si, Ca, Al, Cl, Na, K, S, and Fe. Zn, Mg, and Pb are often found in concentrations exceeding 10,000 mg/kg, while Hg is usually found in concentrations below 10 mg/kg. Other trace elements (i.e., Cu, Sb, Cd, Sr, Ni, As, Ag, Co, V, Mo, and Se) are present in concentrations below 1000 mg/kg.

Table 12.2   Elemental Composition of Bottom Ash from all Types of Incinerators and
Fly Ash, Dry/Semidry, and Wet APC System Residues from Mass Burn
Incinerators

Element	Range for bottom ash (mg/kg)	Range for fly ash (mg/kg)	Range for dry/semidry APC system residues (mg/kg)	Range for wet APC system residue without fly ash (mg/kg)
Ag	0.29–37	2.3–100	0.9–60	—
Al	22,000–73,000	49,000–90,000	12,000–83,000	21,000–39,000
As	0.12–190	37–320	18–530	41–210
B	38–310	—	—	—
Ba	400–3,000	330–3,100	51–14,000	55–1,600
C	10,000–60,000	—	—	—
Ca	37–120,000	74,000–130,000	110,000–350,000	87,000–200,000
Cd	0.3–71	50–450	140–300	150–1,400
Cl	800–4,200	29,000–210,000	62,000–380,000	17,000–51,000
Co	6–350	13–87	4–300	0.5–20
Cr	23–3,200	140–1,100	73–570	80–560
Cu	190–8,200	600–3,200	16–1,700	440–2,400
Fe	4,100–150,000	12,000–44,000	2,600–71,000	20,000–97,000
Hg	0.02–7.8	0.7–30	0.1–51	2.2–2,300
K	750–16,000	22,000–62,000	5,900–40,000	810–8,600
Mg	400–26,000	11,000–19,000	5,100–14,000	19,000–170,000
Mn	83–2,400	800–1,900	200–900	5,000–12,000
Mo	2.5–280	15–150	9.3–29	1.8–44
N	110–900	—	—	1,600
Na	2,900–42,000	15,000–57,000	7,600–29,000	720–3,400
Ni	7–4,300	60–260	19–710	20–310
O	400,000–500,000	—	—	—
P	1,400–6,400	4,800–9,600	1,700–4,600	—
Pb	98–14,000	5,300–26,000	2,500–10,000	3,300–22,000
S	1,000–5,000	11,000–45,000	1,400–25,000	2,700–6,000
Sb	10–430	260–1,100	300–1,100	80–200
Se	0.05–10	0.4–31	0.7–29	—
Si	91,000–310,000	95,000–210,000	36,000–120,000	78,000
Sn	2–380	550–2,000	620–1,400	340–450
Sr	85–1,000	40–640	400–500	5–300
Ti	2,600–9,500	6,800–14,000	700–5,700	1,400–4,300
V	20–120	29–150	8–62	25–86
Zn	610–7,800	9,000–70,000	7,000–20,000	8,100–53,000

Modified from the International Ash Working Group.

The total organic content (TOC) is generally below 10,000 mg/kg for all types of APC residues. The pH for dry/semi-dry APC system residues typically exceeds 12, while wet scrubber sludges have pH values around 10.5. The excess lime in the dry/semi-dry system results in higher buffering capacities compared to electrostatic precipitator (ESP) fly ash and wet sludge.

MWC residues and ashes contain small amounts of polychlorinated dibenzo-p-dioxins/polychlorinated dibenzofurans (PCDDs/PCDFs). In most studies, the levels detected in ashes from modern facilities have been well below levels of concern. In the U.S. EPA/Coalition on Resource Recovery and the Environment

(CORRE) MWC Ash Study, ashes sampled from five modern mass burn facilities contained PCDDs/PCDFs at levels below the Center for Disease Control recommended level of 2, 3, 7, 8–tetrachlorodibenzodioxin (TCDD) Toxicity Equivalency of 1 part per billion in residential soil.[11,12] Note that this level is based on health risk assessments of ingesting soil daily over a lifetime.

LEACHING OF THE RESIDUES

Perhaps the most controversial issue associated with MWC residues is that of leaching or perceived leaching of potentially toxic contaminants, particularly Pb and Cd. Over the years this controversy focused on the fact that at times some of the residues failed the EPA's Extraction Procedure for Toxicity (EP_{TOX}) and later the Toxicity Characteristics Leaching Procedure (TCLP) test used to determine if a waste was hazardous based on "toxicity" characteristics. Even so, these residues were not routinely being managed as hazardous waste because they were considered by many to be exempt from the Resource Conservation and Recovery Act (RCRA) Subtitle C hazardous waste regulations.[13] Some states however have required disposal of the residues into subtitle C-like facilities. This issue eventually led to litigation that was recently decided by the U.S. Supreme Court, which ruled that the residues are not exempt.[14] Technical arguments have been made that laboratory tests such as the TCLP grossly over estimate the actual release of contaminants and do not measure the release rates that one would expect under field conditions. Significant field data supports this argument.

In general, laboratory leaching tests are designed either to simulate a field leaching scenario to measure a specific fundamental leaching property of the material being tested. Specific leaching properties to be measured are availability, solubility, and release rate. Availability is defined as the potential quantity of a species that may leach under the specified conditions over a prolonged period of time. Results of availability tests are reported in units of mass leached/mass initial material (typically mg Pb/kg ash) and are not indicative of the time frame over which release may occur. Availability tests are characterized by high liquid-to-solid ratios (e.g, 100:1), small particle sizes of material to be tested, extreme conditions (e.g., relative to pH), and contact times which permit dissolution to be achieved. Solubility tests are designed to determine the concentration of specific species in solution under saturated conditions. Results of solubility studies are reported in concentration units (typically mg/L). Solubility tests are characterized by low liquid-to-solid ratios (e.g., 2:1), small particles sizes of material to be tested, varied conditions (e.g., relative to pH), and contact times that permit chemical equilibrium to be achieved between the solid phase and solution. Release rate tests are used to estimate the rate of release of specific species from a material. Results are usually reported as a rate per mass or surface area basis (typically mg released/sq material/day or mg released/m^2/day) or for diffusion-controlled release rates as an effective diffusion coefficient. Release rate tests typically are carried out using column flow studies for granular materials and tank leaching studies for monolithic materials. Serial batch tests can also be used to determine release rates over time. Determination of specific leaching properties permits estimation of releases under varied environmental conditions, while employment of leaching tests that simulate a specific disposal scenario do not readily allow transferability of results.

Considerable confusion and debate has resulted from the misinterpretation of leaching test results. For example, the TCLP was designed to simulate codisposal of a material with MSW in a landfill. Thus, results from TCLP testing are not indicative of potential releases under different environmental exposure scenarios. Results of extensive laboratory leaching tests have indicated that the variables that impact release potential of heavy metals the most significantly are solution final pH and liquid-to-solid ratio. Typically, cadmium solubility, and hence release, increases substantially with decreasing pH at pHs less than 8. Lead release increases substantially with decreasing pH at pHs less than 6 and also increases substantially with increasing pH at pHs greater than 11. Thus, with a solution pH between 8 and 11, a minimum release of both lead and cadmium occurs.

The results of synthetic acid rain extractions have not been significantly different from extractions with distilled water because of the high alkalinity of the residuals. Liquid-to-solid ratio impacts most leaching test results because most leaching tests are carried out at a liquid-to-solid ratio where a saturated solution (solubility-controlled) exists at the end of the contacting interval. Thus, greater liquid-to-solid ratios result in release of a greater mass of heavy metal. However, it is important to note that even at extreme laboratory leaching conditions, tests have shown that only a small fraction (typically much less than 10%) of the total heavy metal element present in the residual is released. Conversely, most of the alkali salts present in the residuals are readily available for leaching and are released independent of solution pH or liquid-to-solid ratio.

Field studies of leaching of combined MWC residuals in monofills have indicated that leachate concentrations of heavy metals are overestimated by laboratory leaching tests. Typical leachate concentrations observed for these metals are below or near drinking water standards. However, total dissolved solid (soluble salts) concentrations are several orders of magnitude greater than drinking water standards and, at low liquid to solids ratios, are comparable to sea water.

The EPA/CORRE MWC-Ash Study sampled and analyzed leachate from the ash disposal units of five mass-burn MWC facilities. The data from this study indicated that all the metals were below their EP_{TOX} maximum allowable limit. Also indicated by the data was that some of the metals concentrations in the leachate met primary or secondary drinking water standards and the leachate did not contain significant quantities of PCDDs/PCDFs. Those that are most often found are the highly chlorinated homologs, which are the ones with the relatively lower toxicity equivalency factors (TEFs).[11]

This study also investigated leaching of ashes collected from the facilities with 6 different extraction procedures, including:

1. Acid 1 (EP_{TOX})
2. Acid 2 (TCLP Fluid 1)
3. Acid 3 (TCLP Fluid 2)

4. Deionized water (method SW-924, also the Monofill Waste Extraction Procedure or MWEP)
5. CO_2 saturated deionized water, and
6. Simulated acid rain (SAR).

Analysis showed that the EP_{TOX}, $TCLP_1$, and $TCLP_2$ extracts contained higher levels of the metals than did extracts from the other procedures. While the extracts from the EP_{TOX} and TCLP procedures occasionally exceeded the EP toxicity allowable limit for some metals, extracts from the other three tests did not. The deionized H_2O, the CO_2, and the SAR extractions procedures more closely simulated the concentrations of Pb and Cd in the field leachates.

Some results from five years of leachate sampling and analyses from the North Marion County (Oregon) Disposal Facility (Woodburn Monofill) are summarized below:[12]

- The major leachate constituent is salts and the main salt constituents are chloride, sulfate, calcium, and sodium.
- All metal concentrations in the leachates were below the EP_{TOX} and TCLP maximum allowable limits.
- Iron, manganese, and lead exceeded federal safe drinking standards in leachate from the closed cell; several metals (Al, Ba, Cd, Fe, Pb, Mg, and Hg) exceeded those standards in the leachate from the active cell.
- Salt concentrations in the leachates have generally decreased over the 5-year period. In one exception, the sulfate in the closed cell increased during one sampling year, but then decreased during the last sampling period.
- Levels of pH in the closed cell have remained fairly constant between 6.7 and 7.0. In the active cell, the pH has ranged from 5.7–6.1.

Results from these and similar studies tend to verify technical arguments that the TCLP does over estimate the actual release of metals that one would expect to find in the field.

There are many factors that affect the leaching of constituents from ash. Data compiled by the International Ash Working Group (IWAG) of results from regulatory leaching tests from several countries indicate that concentrations of metals in most of the extracts are solubility-controlled.[10] These results can be modified by changes in ash alkalinity when the final pH of the extraction procedure is not specified. The California WET test is an exception because the metal complexation by citric acid results in availability-controlled leaching.

The approach (proposed by the IAWG and which is a modification of that proposed for the Netherlands) to determine how the ash can ultimately be managed involves several tests. First, total concentrations of ash constituents are determined using appropriate analytical methods. This is followed by an assessment of the amount of these constituents available for leaching under a presumed worst-case condition. This test is referred to as the availability leach test; it determines the maximum quantity of a specific element or species that can be expected to leach. The test uses a high liquid-to-solid ratio (e.g., equal to or greater than 100), strong complexing agents (e.g., citric acid), acidic final pH (e.g., 4 or lower) or, for amphoteric elements, a pH greater than 11. Availability-controlled leaching results should be presented as the measured release (mg element leached/kg of material extracted). The composition of the initial extraction solution, liquid-to-solid ratio, and final pH also should be indicated.

The observed release of specific elements or species of interest under availability-controlled leaching conditions is independent of the conditions used to reach availability-controlled conditions. The following are several experimental conditions that result in availability-controlled release:

- Liquid-to-solid ratios greater than 10, concurrent with pH less than or equal to 4, results in availability-controlled leaching for most elements and species of interest.
- Liquid-to-solid ratios greater than 10, concurrent with near neutral pH and 0.2 molar citric acid, results in availability-controlled leaching for most elements and species of interest.
- Liquid-to-solid ratios greater than 10, concurrent with near neutral pH and 0.016 molar EDTA, results in availability-controlled leaching for most elements and species of interest.

All three of the above availability-controlled leaching conditions result quantitatively in the same release of a specific element or species. The exception is Pb from APC residues, which often shows a maximum availability at pH greater than 11.

The availability of specific elements or species of interest is a fraction of the total concentration present in the ash and may be significantly different for different ash types. Grate siftings are significantly higher in Cu and Pb availability compared to other ash types. APC residues are significantly higher in Cd, Pb, Zn, Cl, Br, and SO_4 availability compared to other types, with the possible exception of Pb in grate siftings. The total soluble mass fraction of APC residues is much greater than that of other ash types.

After determining the amount available for leaching, column leaching tests can be used to assess the release of elements or species as a function of time. Similar tests — such as the Monolith Leach Test, a modification of the ANSI 16.1 test, and the ANSI 16.1 test — can be used to assess the release of the selected species over time.

Based on the information gained from these tests, one can develop estimates of expected releases over time for species of concern when placed in various disposal or utilization options. Cumulative release in the field can be estimated on knowledge of the unified pH curve, availability, anticipated field pH, and the anticipated liquid-to-solid rates. If field redox conditions are also known, estimates can be further refined.

Aging or weathering of ash normally results in a decrease of leachate pH towards neutral. One aging reaction results from uptake of CO_2 and self-neutralization of the ash. These changes result in decreased solubility of many elements and consequently decreased release.

Once a class or type of residues has been completely characterized with respect to leaching behavior and chemical/physical characteristics, then it should only be necessary to apply a good QA/QC program to certify that the residue is the same as others in the class and acceptable for a given utilization option. This or a similar process is not being used in the U.S., except in the case of some utilization demonstrations.

TREATMENT OPTIONS

There are several options available for treating MSW combustion residues. The need for treatment is not clear and depends on site-specific conditions, including regulatory requirements and utilization or disposal objectives. Treatment options include:

Ferrous Metal Recovery — Magnetic separation and screening of ferrous metal from the bottom ash is routinely practiced at many WTE facilities. Ferrous metal accounts for approximately 15% of the bottom ash (from mass burn facilities) and is normally sold as scrap. The ferrous metal content of the bottom ash from RDF combustion facilities may be less due to more extensive processing of the MSW prior to combustion. In some cases, separation of nonferrous metals by eddy current separations may also be practiced.

Compaction — The physical properties of the residues are such that compaction at optimum moisture content can be effective in significantly reducing the permeability of the residues when placed in a monofill or other disposal/utilization options. This reduces infiltration of water and, therefore, the potential for release of contaminants.

Tests have shown that bottom ash can be compacted to Proctor densities of 1,600 kg/M^3 or more. At optimum moisture content of approximately 16% and optimum compaction effort, saturated hydraulic conductivities are around 10^{-6} cm/sec. Depending on ash characteristics, optimum moisture contents and compactive effort may change. Laboratory and field permeabilities of less than 10^{-8} cm/sec have been reported. Compacted bottom ash exhibits good bearing capacities as measured by the California Bearing Ratio Test.

Classification — Screening or separation of ashes into oversized and undersized fractions can be beneficial and is in some cases necessary for utilization purposes. Screening coarser material (3/4-inch or perhaps 3/8-inch) from bottom ash improves the characteristics as a coarse aggregate. Bottom ash used in the Laconia, NH asphalt paving project was screened to less than 3/4-inch. This ash also did not contain grate siftings or heat recovery ash and had been aged for 5 months.[9] From a physical perspective, fly ash can be screened to meet specifications for fine cement aggregate (ASTM C33). The size distribution of the fly ash may also make it suitable for use as a soil aggregate for paving applications based on ASTM Standard 1241.[15]

Solidification/Stabilization (S/S) — S/S refers to technologies or processes that use additives (or binders) to physically and/or chemically immobilize hazardous constituents in wastes, soils, and sludges.[16,17] Solidification normally refers to the conversion of a liquid or semi-liquid to a solid form. While solidification does not necessarily involve a chemical interaction of the constituent of concern with the solidifying agent, the process can restrict contaminant mobility by encapsulating the contaminant in a treated product of reduced surface area, lower permeability and better handling characteristics. Stabilization converts the contaminants into less soluble or less toxic forms without necessarily achieving solidification. The best approach to S/S normally is to chemically stabilize, then solidify the waste. The most common binders used are inorganic systems based on cement and pozzolanic materials, although some hazardous waste S/S processes have used thermoplastic binders, synthetic polymers and organophylic clays. The process involves mixing binders with the waste and water, sometimes incorporating additives such as sodium silicate, cement kiln dust, coal fly ash, bentonite or proprietary materials. The ratio of waste, binder, other additives and water should be tested to optimize the final product integrity and performance (contaminant leachability, strength, curing rate, etc.) to meet required treatment objectives. Depending on the S/S process, the final product may be either monolithic or granular in nature.

S/S processes using Portland cement and similar binders will usually result in a treated product with more weight and increased volume. This could significantly affect transportation and disposal costs. The sequence of binder additives and mixing may significantly affect the quality of the product and its ultimate performance. Adequate mixing is also important.

S/S processes are not effective for treating soluble salts. Therefore, since APC residues contain high levels of salts, leaching of these salts from the S/S matrix will likely result in poor performance. The subsequent loss of physical properties and durability of the treated product may result in increased release of the metals.

The addition of Portland cement and other additives is being practiced by several facilities in the U.S. One example is the Commerce Refuse-to-Energy Facility (CREF) located in Commerce, CA.[18] The treatment facility consist of bottom ash conveying, separation and storage equipment; a fly ash conveyor and silo; and treatment equipment similar to a concrete batching plant. The treatment process mixes screened bottom ash with fly ash, Type II Portland cement and water. A standard concrete mixer truck is used to mix the batch. The slurry is discharged into roll-off-bins. After setting, the resulting approximate 16-ton monolith is transported to a non-hazardous waste landfill where it is crushed and used as road sub-base aggregate. The bottom ash larger than 2″ in size is removed by screening prior to treatment.

The WES-Phix® Ash Immobilization Process, patented by Wheelabrator Environmental Systems Inc., can be classified as an S/S process. The process uses water-soluble phosphates and alkali to reduce metals solubility by the formation of insoluble (or less soluble) mineral phases. Treatment by the phosphate may result in insoluble phosphate compounds of lead, copper, and zinc. The reduction of Pb leachability was confirmed by tests conducted by the EPA in an evaluation of S/S processes for treating MSW combustion residues,[4,19] This evaluation did not confirm that Cd was chemically treated. The WES-Phix process is being used at several facilities in the U.S. Rights to use the process are available from Wheelabrator.

S/S processes can be effective in treating MWC residues. However, several factors must be carefully considered.

- The high concentration and ultimate fate of soluble salts must be carefully considered in the design of the process. Pretreatment to remove these may be required.
- Evaluation of the S/S process design, performance, and treatment efficiency should be based on a matrix of several testing protocols. A single test such as the TCLP cannot provide the information required to completely evaluate the potential for release of contaminates or physical durability.
- Many of the available S/S processes, particularly those using binders such as Portland cement, will increase volume. This could result in added transportation and disposal costs.

Chemical Extraction — Processes for chemically extracting metals have been researched and developed at the laboratory scale.[20-22] Chemical extraction has not been practiced on a commercial scale primarily because of economics. Extraction of soluble salts has also been demonstrated in the laboratory and is less expensive, but also, has not been implemented in the U.S.

Vitrification and other Thermal Processes — Vitrification results in the melting of the ash with glass forming additives to incorporate the contaminants into an alumina-silicate matrix. The process can substantially reduce the volume of materials 60% or more and usually results in a product more resistant to leaching, however there are several concerns. Costs can be high, ranging from $100 to $200 per ton. Release of contaminants during melting (e.g., organics and volatization of some metals) may require additional air emissions control and subsequent treatment of the collected residues. Vitrification, while researched and demonstrated in the U.S., has not been implemented for treating MWC residues.[22-25]

Heating of the residues can also cause sintering, mineral respeciation or melting to "fuse" the material into a slag. Bottom ash fusion and perhaps vitrification is practiced at several facilities in Japan, but not in the U.S.

The necessity for treating residues depends on several factors. These include local regulatory requirements, disposal objectives, specifications required for utilization, liability considerations, economics, and other site-specific factors.

DISPOSAL

The prevailing method in the U.S. for managing combined residues is by monofill disposal. The monofills are lined with clay, synthetic liners, or a combination of both and have provisions for leachate collection and treatment. Even before the recent Supreme Court decision which ruled that residues were not exempt from RCRA Subtitle C hazardous waste regulations, some states required testing of ashes to determine if they should be classified as hazardous. Some states already required disposal into landfills very similar in design to subtitle C landfills. In some instances, they were managed as a hazardous waste if they failed these tests. The ruling now means that owners of WTE facilities must determine if ashes are hazardous. If they are, owners will be required to manage the ash as a hazardous waste. This could result in ultimate disposal into a Class C hazardous waste landfill and added costs. On the other hand, if the ashes are determined not to be hazardous, they could be classified as a subtitle "D" non-hazardous waste and disposed into Subtitle D, part 258, municipal solid waste landfills. Some states, however, may still require more stringent requirements.

In some countries residues are separated into bottom ash and APC residues prior to management. The APC residues are treated, usually with S/S techniques, prior to disposal into landfills. In Germany APC residues may be stored in salt mines. In some instances in Denmark ashes have been placed in fills near the ocean and allowed to leach into the more saline ocean waters. Some countries are also considering the concept of "controlled contaminant release." The controlled release concept implies that residues would be managed in a manner that controls release rates and the quality of the leachate within preestablished levels that are consistent with surrounding conditions. The leachate is allowed to discharge into the surroundings as it is formed as long as it does not exceed preestablished acceptable levels. Siting is critical. This concept is being considered by some countries as a result of concern that "dry entombment" in lined and capped landfills will result in greater uncontrolled release of contaminants in the future. Controlled release is not being used or considered, at least on a regulatory basis, in the U.S.

UTILIZATION

In some European countries 50% or more of the bottom ash is utilized, but utilization of MWC residues is not routinely practiced in the U.S. There has, however, been significant interest in evaluation of utilization alternatives including a number of research and demonstration projects. These projects include asphalt pavement, construction blocks, artificial reefs, shoreline erosion control, and similar applications. Research to improve fundamental knowledge of how residues perform in alternative situations include projects to determine the elemental speciation and ultimate life of constituents in untreated and treated residues;[26] mobility of dioxins and furans from stabilized residue in seawater;[27] and environmental evaluation of a boathouse constructed of blocks containing MWC residues.[28] Results from these types of projects and from actual field demonstrations are providing data critical in evaluating, designing, and implementing utilization options to protect the environment and human health.

Primary applications under consideration in the U.S. include the use of bottom or combined ash as an aggregate substitute in Portland cement, as bituminous concrete (pavement), as sub-base in roads, as fill material under contained conditions such as covered embankments, and as daily cover in landfills. Marine applications (e.g., shore erosion control and artificial reefs) are also under consideration.

The following summarizes MWC utilization status in other countries:

Canada — Utilization is not currently being considered.

Germany — Approximately 50% of bottom ash is utilized primarily as road bases and sound barriers on autobahns. There are also pilot-scale tests being conducted to evaluate APC residue as grout in coal mines. Another consideration involves use of APC residue to form alinite cement.

The Netherlands — The goal in The Netherlands is to utilize at least 80% of the residues. Currently 40% of the fly ash produced is used as a fine aggregate in asphalt. More than 2,000,000 tons of bottom ash (approximately 60%) have been used as road base, embankments, aggregate in concrete, and aggregate in asphaltic concrete. To reach this goal, however, The Netherlands has established a policy and regulatory framework which establishes standard leaching tests, composition requirements, evaluation protocols for building materials, and certification of residues for utilization.

Denmark — Utilization of bottom ash was initiated in Denmark in 1974. Approximately 72% of the bottom ash, including grate siftings and heat recovery ash, screened, is used as sub-base at parking lots, bicycle paths, and paved and unpaved roads. Requirements for ash composition for utilization are pH > 9.0 in a 1% slurry, alkalinity greater than 1.5 eq/kg, Pb content below 3,000 mg/kg, and Hg content less than 0.5 mg/kg, as determined by HNO_3 digestion. In addition, there are restrictions which include a 20 m minimum distance to water wells, 1–2 m maximum thickness under a pavement, and a maximum of 200 m³ and 0.30 m maximum thickness for unpaved applications.

Sweden — Sweden has adopted a policy that utilization should improve the general environmental conditions and create a smaller environmental impact than disposal. Regulations for MWC ash utilization are currently being developed.

Based on technical data and findings by the IAWG and others, the following factors should be considered in selecting, designing, and implementing MWC residue utilization:

- Bottom ash can be effectively utilized as an aggregate substitute in several applications if appropriate engineering (structural) criteria and environmental performance guidelines are met. Large-scale utilization projects processed as structural soil in Europe include its use in land reclamation from the ocean, wind barriers, sound barriers, road sub-bases, and parking lot and bike path bases. Demonstrations in the U.S. have bottom ash as aggregate substitutes in bituminous pavements, sometimes at high substitution rates. In other demonstrations, bottom ash has been used as an aggregate substitute in Portland cement applications for marine reefs, shore protection devices, and service buildings.
- The high alkalinity caused by the addition of cement in S/S applications influences the release of some metals in a negative manner. This must be considered in the design.
- Use of bottom ash as structural fill in large applications such as embankments may lead to high salt loads and relatively high leachability of salts. Pretreatment to remove soluble salts may be necessary in some applications.

- The use of bottom ash should be investigated in stabilized road-based applications using cold-emulsion bituminous materials, and as final cover at landfills.
- Ferrous and non-ferrous metals should be separated from bottom ash prior to use as an aggregate.
- Bottom ash should be stockpiled prior to utilization for an estimated one to three months with adequate moisture so swelling, hydration, carbonation, and oxidation reactions can occur. This benefits structural durability and chemical stability. Appropriate methods should be used to control and manage any leachate and/or runoff during stockpiling.
- The design, construction, and implementation of any MWC residue utilization project should be based on a complete understanding and knowledge of the ash characteristics, the ash product behavior in the situation, environmental conditions, physical requirements, regulatory requirements, and similar factors. This will necessitate appropriate testing and evaluation protocols, including tests to determine elemental concentration, availability of trace metals and total soluble salts for release, estimates of contaminant release over time and under field conditions, physical durability, and similar factors. This cannot be done by relying on a single test such as the TCLP.
- The acceptability of a product for utilization should be based on the amount of anticipated contaminant release (assuming that physical and other specifications are met) over the expected product lifetime. This should take into consideration both contaminant release potential and release rates. The acceptable level of contaminant release should be defined for each utilization — can be based on release rates, the potential of similar products from natural materials, and environmental considerations.
- The time frame of intended use should be considered and an analysis conducted on the ultimate fate of the material following completion of the specified use period. Many of the utilization scenarios for bottom ash have release rates or release potentials similar to natural materials.
- Granular material (e.g., roadbase) will behave differently than monolithic applications and require different testing protocols (e.g., compacted granular leach test). Monolithic ash products should be evaluated using appropriate monolithic leach tests (e.g., tank leaching) and also for durability.
- Appropriate quality control and quality assurance procedures should be developed and implemented to maintain ash product quality.

There are a number of other factors to consider, such as distance to sensitive natural areas, local requirements, monitoring during the product use and similar factors.

Although not routinely practiced in the U.S., mounting evidence supports the technical argument for separating fly ash from bottom ash and further separation of grate siftings from bottom ash. While this may not be necessary for disposal in lined monofills, as evidenced by the quality of leachates from several disposal facilities, separation definitely appears valid for most potential utilization scenarios. Grate siftings are greatly enriched in the fraction of Pb most available for leaching and Al, which can cause hydrogen generation, subsequent swelling, and reduced physical durability.

The APC residues contain relatively high levels of soluble salts and trace metals. Soluble salts are difficult to treat and they readily leach from the matrix.

SUMMARY

Combustion to recover energy and reduce the volume of waste that requires landfilling is an important factor in the management of MSW in the U.S. This process is applied to approximately 16% of the MSW generated. In other countries, from 30–60% is combusted. This generates residues that must be managed in ways that protect human health and the environment.

MSW residues contain varying amounts of trace elements that raise questions covering their classification and management. Cadmium and lead represent chief concerns, although soluble salts could also be problematic if not properly managed. Some MWC ashes have failed regulatory leach tests such as the TCLP, primarily for Pb and Cd. Technical arguments that these tests do not mimic field conditions are supported by analyses of leachates from ash monofills, which have shown levels of trace metals below the TCLP regulatory limits and, in significant cases, below drinking water standards. The major constituent in the leachates from these sites has been salts.

Residues are primarily managed in the U.S. by disposal into lined monofills. Their utilization is not routinely practiced in the U.S. Opportunities and technical bases exists, however, for utilization. Utilization options being considered and demonstrated include aggregate for fill, aggregate in asphaltic applications, concrete applications (e.g., shoreline erosion control, construction blocks), landfill daily cover, and others. Technical data suggest that APC residues and grate siftings should be collected separately from bottom residues. Grate siftings contain enriched levels of Pb, with almost 50% of the Pb available for leaching, and Al, which may result in hydrogen generation. APC residues are enriched in trace metals and contain the largest fraction of leachable Cd. They also have high concentrations of soluble salts.

Residues for utilization should be thoroughly evaluated for constituent concentrations, concentrations of species of concern that are available for leaching, release rates over time, and expected contaminate releases over time for the utilization option under consideration. Protocols are available to determine these factors. One cannot rely on a single test such as the TCLP to make these assessments.

Based on their technical aspects, processed bottom residues should be considered for utilization. Research and demonstrations are providing data that will help verify proper utilization in a manner protective of human health and the environment. The ultimate affect of the Supreme Court's decision on ash management and particularly ash utilization remains unclear. However, certain ashes will now have a legal basis to be classified as not hazardous. This could help alleviate some objections to ash utilization. Utilization, however, must follow sound scientific and engineering principles and be conducted with appropriate measures to assure that it is acceptable to the environment and to human health.

ACKNOWLEDGMENTS

Much of the material presented in this chapter was adopted from information provided by the International Ash Working Group (IAWG). Members of the IAWG are:

Han van der Sloot, Ph.D., Chairman
Netherlands Energy Research
  Foundation
Petten
The Netherlands

John Chandler
Chandler and Associates, Inc.
Willowdale, Ontario
Canada

Taylor Eighmy
University of New Hampshire
Durham, New Hampshire
United States of America

Jan Hartlén
Swedish Geotechnical Institute
Linköping
Sweden

Steve Sawell, Project Manager
Compass Environmental
Burlington, Ontario
Canada

Ole Hjelmar
Danish Water Quality Institute
Horsholm
Denmark

David S. Kosson
Rutgers, The State University of New
  Jersey
New Brunswick, New Jersey
United States of America

Jürgen Vehlow
Kernforchungs Zentrum Karlsruhe
Karlsruhe
Germany

REFERENCES

1. Kiser, J.V.L., Update of U.S. Plants, *The IWSA Municipal Waste Combustion Directory: 1993*, Integrated Waste Services Association, Washington, D.C., 1993.
2. Energy-from-Waste, 1993 Activity Report, *Solid Waste & Power*, HCI Publications, Kansas City, MO, 1993.
3. Toxicity Characteristics Leaching Procedure (TCLP), *Federal Register*, November 7, 1986, 51 (216), pp. 40643–40652.
4. Kosson, D.S., Kosson, T., and van der Sloot, H., Evaluation of Solidification/Stabilization Treatment Processes for Municipal Waste Combustion Residues, EPA/600/SR-93/167, Risk Reduction Engineering Laboratory, U.S. EPA, Cincinnati, OH, October 1993.
5. Wiles, C.C., Issues Associated with Municipal Waste Combustion Residue Utilization and Results of U.S. EPA Research, *Proceedings National Solid Waste Forum, July 1992*, Association of State and Territorial Solid Waste Management Officials, Portland, OR.
6. Kosson, D.S., Clay, B.A., van der Sloot, H.A., and Kosson, T.T., Utilization Status, Issues and Criteria Development for Municipal Waste Combustion Residues in the U.S., *Environmental Aspects of Construction with Waste Materials*, Eds., Goumans, J.J.J.M., van der Sloot, H.A., and Aalbers, Th.G., Elsevier Science B.V., Amsterdam, The Netherlands, 1994.

7. Kosson, D.S. and Hoffman, F.E., Management of Municipal Waste Combustor Residues: Issues and Options; *Proceedings of the International Power Generation Conference, October 1992*, American Society of Mechanical Engineers, Atlanta, GA.

8. Chesner, W.H. and Roethal, F.J., Eds., *Proceedings of the Fifth International Conference on Municipal Solid Waste Combustion Ash Utilization*, Arlington, VA, November 1992.

9. Musseleman, C. N., Eighmy, T. T., Gress, D. L., Killeen, M. P., Presher, J. R., and Sills, M. H., The New Hampshire Bottom Ash Paving Demonstration, U.S. Route 3, Laconia, New Hampshire, *Proceedings of the 16th Biennial 1994 National Waste Processing Conference,* ASME, New York, NY, 1994, 83–90.

10. An International Perspective on Municipal Waste Incineration Residue Characterization and Management, International Conference on Environmental Implications of Construction Materials and Technology Developments, Maastricht, The Netherlands, 1–3 June 1994, The International Ash Working Group (IAWG).

11. Roffman, H.K, Major Findings of the U.S. EPA/CORRE MWC-Ash Study, *Proceedings Municipal Waste Combustion Conference Papers and Abstracts from The Second Annual International Specialty Conference*, Air & Waste Management Association; Tampa, FL, April 1991, pp. 96–123.

12. Roffman, H.K., Results of Fifth Year Leachate and Ash Aging Study and Plans for the Future Woodburn Monofill Site, *Proceedings of the Sixth International Conference on Municipal Solid Waste Combustor Ash Utilization,* November 16–17, 1993, Resource Recovery Report, Arlington, VA, 1994.

13. Resource Conservation and Recovery Act of 1976 (RCRA), PL 94-580, October 21, 1976 as amended through PL 102-389, October 6, 1992.

14. U.S. Supreme Court, [May 2, 1994]. *City of Chicago, et al., Petitioners v. Environmental Defense Fund et. al.,* 114 S. Ct. 1588.

15. Goodwin, R.W., Defending the Character of Ash, *Solid Waste & Power*, HCI Publications, Kansas City, MO, September/October 1992.

16. Wiles, C.C., (1987), A Review of Solidification/Stabilization Technology, *Journal of Hazardous Materials,* 14, 5, 1987.

17. Conner, J.R., *Chemical Fixation and Solidification of Hazardous Wastes*, Van Nostrand Reinhold, New York, 1990.

18. Eaton, M. and Korn, J., The Commerce Refuse-To-Energy Facility Ash Handling and Treatment System, *Proceedings of the Fifth International Conference on Municipal Solid Waste Combustor Ash Utilization,* November 17–18, 1992, Arlington, VA, pp. 141–154.

19. Wiles, C.C., The U.S. EPA Program for Evaluation of Treatment and Utilization; Air Technologies for Municipal Waste Combustion Residues, *Proceedings of Municipal Waste Combustion*, April 1991, Air & Waste Management Association, Tampa, FL.

20. Legiec, I.A. and Kosson, D.S., Recovery of Heavy Metals from MSW Residues; Pilot Plant Design and Operations, *AIChE 1989 Summer National Meeting*, Philadelphia, PA, August 1989.

21. Legiec, I.A., (May 1991), *Design and Scale-Up of a Heavy Metals Recovery From Incinerator Ashes* (B.S. and M.S. Rutgers, the State University of New Jersey, Graduate Assistant, 1989 J. Lagrosa Award NJWPCA).

22. Solid Waste & Power staff, Treatment and Use: The Future of Ash Management, *Solid Waste & Power*, HCI Publications, Kansas City, MO, October 1991.

23. Kosson, T., Kosson, D.S., Stuart, B., Wexell, D., and Stempin, J., Vitrification of Municipal Solid Waste Combustion Air Pollution Control Residues Using Corning, Inc. Process, *Municipal Waste Combustion, Proceedings of International Specialty Conference,* March 1993, Williamsburg, VA, Air & Waste Management Association, Pittsburgh, PA, VIP-32, 1993, pp.784–794.
24. Oden, L.L. and O'Connor, W.K., ASME/US Bureau of Mines Investigative Program Report on Vitrification of Residue (Ash) from Municipal Waste Combustion Systems, The American Society of Mechanical Engineers, CRTD-Vol. 24, United Engineering Center, New York, NY, 1994.
25. Wexell, D., Vitrification of MSW Fly Ash for Heavy Metal Stabilization, *4th Annual ACS Division of Industrial Engineering Technology for Hazardous Waste Management, September 21–23, 1992,* Atlanta, GA.
26. Eighmy, T.T., Domingo, D., Krazanowski, J.R., Stuämpfli, D., and Eusdem, D., The Speciation of Elements in MSW Incinerator Residues, *Municipal Waste Combustion, Proceedings of the International Specialty Conference, March 1993, Williamsburg, VA,* Air and Waste Management Association, Pittsburgh, PA, VIP-32.
27. Roethel, F.J. and Breslin, V.T. (1995), Behavior of Dioxin/Furans and Metals Associated with Stabilized MSW Combuster Ash in Seawater, *Chemistry in Ecology,* 10, 3/4, 259.
28. Roethel, F. and Breslin, V.T., Stoney Brook MSW Combustor Ash Demonstration Programs, *Proceedings of the Third International Conference on Municipal Solid Waste Combustor Ash Utilization,* November 13–14, 1990, pp. 237–257.

BIBLIOGRAPHY

Sawell, S.E., Chandler, A.J., Eighmy, T.T., Hartlén, J., Hjelmar, O., Kosson, D., Van der Sloot, H.A., and Vehlow, J., The International Ash Working Group: A Treatise on Residues from MSW Incinerators, *Environmental Aspects of Construction with Waste Materials,* Goumans, J.J.J., Van der Sloot, H.A., and Aalbers, Th.G., Eds., Elsevier Science B.V., the Netherlands, 1994, pp. 3–6.
Wiles, C.C., Emphasis of Research by U.S. EPA on Municipal Waste Combustion Residues, *Proceedings from the 3rd International Conference on Municipal Solid Waste Combustion,* Williamsburg, VA, 1993.

CHAPTER 13

Construction Quality Assurance/Quality Control for Landfills

David E. Daniel and Robert M. Koerner

INTRODUCTION

This chapter discusses procedures for quality assurance (QA) and quality control (QC) for municipal solid waste landfills and includes a discussion of principles and concepts. It is not possible to discuss each of the QA/QC procedures for all of the numerous possible components of a landfill. The chapter focuses on two of the most critical components: compacted soil liners and geomembrane liners. Further details on QA/QC for clay and geomembrane liners, plus details of QA/QC recommendations for other components of a landfill, may be found in a recently completed technical guidance document prepared by the authors for the U.S. Environmental Protection Agency.[1]

Construction quality assurance (CQA) and construction quality control (CQC) are widely recognized as critical factors in overall quality management for landfills. The best of designs and regulatory requirements will not necessarily translate to landfills that are protective of human health and the environment unless the waste containment and closure facilities are properly constructed. Additionally, for geosynthetic materials, e.g., geomembrane liners, manufacturing quality assurance (MQA), and manufacturing quality control (MQC) are equally important.

The following definitions are made:

- *Construction Quality Control (CQC)*: A planned system of inspections that is used to directly monitor and control the quality of a construction project. Construction quality control is normally performed by the geosynthetics installer, or for natural soil materials by the earthwork contractor, and is necessary to achieve quality in the constructed or installed system.

- *Construction Quality Assurance (CQA)*: A planned system of activities that provides the owner and permitting agency assurance that the facility was constructed as specified in the design. Construction quality assurance includes inspections, verifications, audits, and evaluations of materials and workmanship necessary to determine and document the quality of the constructed facility.

Responsibilities

The principal organizations and individuals involved in designing, permitting, constructing, and inspecting a landfill are:

- *Permitting Agency.* The permitting agency, which is often a state regulatory agency, is responsible for reviewing the owner/operator's permit application, including the site-specific MQA/CQA plan, for determining compliance with the agency's regulations. The permitting agency also has the responsibility to review all MQA/CQA documentation during or after construction of a facility, possibly including visits to the manufacturing facility and construction site to observe the MQC/CQC and MQA/CQA practices, to confirm that the approved MQA/CQA plan was followed and that the facility was constructed as specified.
- *Owner/Operator.* This is the organization that will own and operate the disposal unit. The owner/operator is responsible for the design, construction, and operation of the landfill. This responsibility includes complying with the requirements of the permitting agency, the submission of MQA/CQA documentation, and assuring the permitting agency that the facility was constructed as specified in the construction plans and specifications and as approved by the permitting agency.
- *Owner's Representative.* The owner/operator usually has an official representative who is responsible for coordinating schedules, meetings, and field activities.
- *Design Engineer.* The design engineer's primary responsibility is to design a landfill that fulfills the operational requirements of the owner/operator, complies with accepted design practices for landfills, and meets or exceeds the minimum requirements of the permitting agency. The design engineer may be requested to change some aspects of the design if unexpected conditions are encountered during construction (e.g., a change in site conditions, unanticipated logistical problems during construction, or lack of availability of certain materials). Because design changes during construction are not uncommon, the design engineer is often involved in the MQA/CQA process.
- *Manufacturer.* Many components, including all geosynthetics, of a waste containment facility are manufactured materials. If requested, the manufacturer should provide information that describes the quality control (MQC) steps that are taken during the manufacturing of the product. The owner/operator, permitting agency, design engineer, fabricator, installer, or MQA engineer may request that they be allowed to observe the manufacture and quality control of some or all of the raw materials and final product to be utilized on a particular job; the manufacturer should be willing to accommodate such requests.
- *General Contractor.* The general contractor has overall responsibility for construction of a waste containment facility and for CQC during construction. The general contractor arranges for purchase of materials that meet specifications, enters into a contract with one or more fabricators (if fabricated materials are

needed) to supply those materials, contracts with an installer (if separate from the general contractor's organization), and has overall control over the construction operations, including scheduling and CQC. The general contractor has the primary responsibility for ensuring that a facility is constructed in accord with the plans and specifications that have been developed by the design engineer.

- *Installation Contractor.* Manufactured products (such as geosynthetics) are placed and installed in the field by an installation contractor who is the general contractor, a subcontractor to the general contractor, or is a specialty contractor hired directly by the owner/operator. The installer is responsible for handling, storage, placement, and installation of manufactured and/or fabricated materials. The installer should have a CQC plan to detail the proper manner that materials are handled, stored, placed, and installed.

- *Earthwork Contractor.* The earthwork contractor is responsible for grading the site to elevations and grades shown on the plans and for constructing earthen components of the waste containment facility, e.g., compacted clay liners and granular drainage layers according to the specifications. The earthwork contractor should have a CQC plan (or agree to one written by others) and is responsible for CQC operations aimed at controlling materials and placement of those materials to conform with project specifications.

- *CQC Personnel.* Construction quality control personnel are individuals who work for the general contractor, installation contractor, or earthwork contractor and whose job it is to ensure that construction is taking place in accord with the plans and specifications approved by the permitting agency.

- *MQA/CQA Engineer.* The MQA/CQA engineer has overall responsibility for manufacturing quality assurance and construction quality assurance. The engineer is usually an individual experienced in a variety of activities, although particular specialists in soil placement, polymeric materials, and geosynthetic placement will invariably be involved in a project. The MQA/CQA engineer is responsible for reviewing the MQA/CQA plan as well as general plans and specifications for the project so that the MQA/CQA plan can be implemented with no contradictions or unresolved discrepancies. Other responsibilities of the MQA/CQA engineer include education of inspection personnel on MQA/CQA requirements and procedures and special steps that are needed on a particular project, scheduling and coordinating of MQA/CQA activities, ensuring that proper procedures are followed, ensuring that testing laboratories are conforming to MQA/CQA requirements and procedures, ensuring that sample custody procedures are followed, confirming that test data are accurately reported and that test data are maintained for later reporting, and preparation of periodic reports. The MQA/CQA engineer is usually the MQA/CQA certification engineer who certifies the completed project.

- *MQA/CQA Personnel.* Manufacturing quality assurance and construction quality assurance personnel are responsible for making observations and performing field tests to ensure that a facility is constructed in accord with the plans and specifications approved by the permitting agency. MQA/CQA personnel normally are employed by the same firm as the MQA/CQA engineer, or by a firm hired by the firm employing the MQA/CQA engineer. Construction MQA/CQA personnel report to the MQA/CQA engineer.

- *Testing Laboratory.* Many MQC/CQC and MQA/CQA tests are performed by commercial laboratories. The testing laboratory should have its own internal QC plan to ensure that laboratory procedures conform to the appropriate American Society for Testing and Materials (ASTM) standards or other applicable testing standards. The testing laboratory is responsible for ensuring that tests are performed in accordance with applicable methods and standards, for following internal QC procedures, for maintaining sample chain-of-custody records, and for reporting data.

- *MQA/CQA Certifying Engineer.* The MQA/CQA certifying engineer is responsible for certifying to the owner/operator and permitting agency that, in his or her opinion, the facility has been constructed in accord with plans and specifications and MQA/CQA document approved by the permitting agency. The certification statement is normally accompanied by a final MQA/CQA report that contains all the appropriate documentation, including daily observation reports, sampling locations, test results, drawings of record or sketches, and other relevant data.

The interaction of the various organizations is summarized in Figure 13.1.

Quality Assurance Plan

Quality assurance begins with a written quality assurance plan that precedes any field construction activities. The MQA/CQA plan should include a detailed description of all MQA/CQA activities that will be used during materials manufacturing and construction to manage the installed quality of the facility. Most state and federal regulatory agencies require that a written MQA/CQA plan be submitted by the owner/operator and be approved by that agency prior to construction. The MQA/CQA plan is usually part of the permit application.

Meetings

Communication is extremely important to quality management. Quality construction is easiest to achieve when all parties involved understand clearly their responsibility and authority. Meetings can be very helpful to make sure that responsibility and authority of each organization is clearly understood. During construction, meetings can help to resolve problems or misunderstandings and to find solutions to unanticipated problems that have developed. Meetings should include:

- *Pre-Bid Meeting.* The first meeting is held to discuss the MQA/CQA plan and to resolve differences of opinion before the project is let for bidding. The pre-bid meeting is held before construction bids are prepared so that the companies bidding on the construction will better understand the level of MQC/CQC and MQA/CQA to be employed on the project.
- *Pre-Construction Meeting.* The pre-construction meeting is held after a project manager is identified and/or a general construction contract has been awarded and the major subcontractors and material suppliers are established. It is usually held concurrent with the initiation of construction. The purpose of this meeting is to review the details of the MQA/CQA plan, to make sure that the responsibility and authority of each individual is clearly understood, to agree on procedures to resolve construction problems, and to establish a foundation of cooperation in quality management.

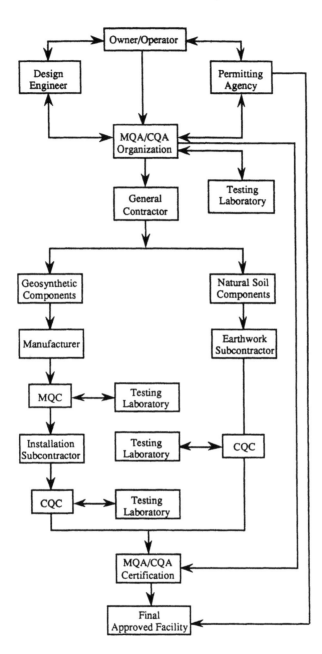

Figure 13.1   Organizational plan for construction quality control and assurance. (From Daniel, D.E. and Koerner, R.M., *Waste Containment Facilities, Guidance for Construction, Quality Assurance, and Quality Control of Liner and Cover Systems,* American Society of Civil Engineers, New York, 1995. With permission.)

- *Progress Meetings.* Weekly progress meetings should be held. Weekly meetings can be helpful in maintaining lines of communication, resolving problems, identifying action items, and improving overall quality management. At all times the MQA/CQA engineer, or designated representative, should be present.

Critical Elements

Construction quality assurance typically focuses on three primary elements of a landfill:

- *Compacted Clay Liners.* The liner system for a municipal solid waste landfill will often contain a low-permeability, compacted clay liner (CCL) that has a typical thickness of 600 mm (2 ft) and a hydraulic conductivity that is less than or equal to 1×10^{-7} cm/s. The materials used in constructing a compacted clay liner should be inspected continuously during handling, and tests of various types should be performed at prescribed frequencies set forth in the CQA plan to verify that the materials are suitable. Compaction of the soil should also be inspected continuously, and various tests are normally part of the process of verifying that the materials were placed and compacted as described in the specifications for the project. Particular care must be taken not to allow the clay liner to desiccate or freeze either during or after construction nor to allow isolated stone/gravel particles to protrude from the surface.
- *Geomembranes.* The geomembrane must be manufactured to stringent and verifiable standards — MQC is critical to production of quality liners that will not be subject to unanticipated degradation. Proper handling and placement of geomembranes is essential to avoid damage to the thin liner materials. Perhaps no aspect of CQA for landfills is more critical than the field seaming of geomembrane rolls or panels. A strictly-followed procedure for CQC, including destructive tests on test strips, and CQA, including destructive and non-destructive tests on seams, is essential. Also, the placement of the layer of material that will overlie a geomembrane is critical: the placement operation must not cause damage to the geomembrane.
- *Soil Drainage Materials.* Most landfills with liners also contain one or more drainage layers that may be constructed of soil materials such as sand and gravel. The percentage of fine material in the drainage system is perhaps the single most critical construction variable: if too much fine material is present, the soil will fail to achieve the desired high hydraulic conductivity.

In addition, other elements are often important:

- *Geosynthetic Clay Liners.* Geosynthetic clay liners (GCLs), which are sometimes used in liner and cover systems for municipal waste landfills, are thin layers of bentonite that are frequently used to augment or in some cases replace a compacted clay liner. Because GCLs are manufactured materials, they are subject to the usual requirements for carefully documented MQC/MQA. Critical CQC/CQA steps are observation of subgrade conditions, overlap of GCL panels, and placement of the overlying materials.
- *Geosynthetic Drainage Materials.* Geosynthetic drainage systems include geonets, geotextiles, and geocomposites. Because these are manufactured materials, there should be provisions for MQC by the manufacturer of these materials and for MQA. If the geosytnthetic is to be used for drainage, the in-plane

transmissivity is critical, and the CQC/CQA process should ensure that the materials are placed in a manner that does not compromise this ability. Geotextiles are frequently used for filters, and they must meet specifications to ensure adequate filtration capability as well as be installed properly, with proper sewing of seams and without damage to the material or excessive exposure to ultraviolet radiation.

- *Ancillary Materials*. Pipes, sumps, anchor trenches, and other details within the landfill are sometimes the most critical points in the landfill and warrant special and detailed CQA, both in terms of materials (including MQC/MQA) as well as installation and protection.

Final Certification

The final step in the CQA process is a document that describes the CQA process, including tests and observations, certification by the certifying CQA engineer, and issuance of a permit to dispose of waste by the permitting agency.

Scope of Chapter

Because of space limitations, this chapter will focus on just two primary elements of a landfill: the compacted clay liner and the geomembrane liner. Detailed information on the other elements is provided by Daniel and Koerner.[1] The coverage here is not intended to be complete but, instead, is intended to give the reader a clear picture of the level of detail that is needed for first-rate QA/QC. The authors have intentionally chosen to focus on a natural material used in a landfill (compacted clay liner) and a manufactured material (geomembrane) to contrast the differing nature of the requirements for these two types of materials.

COMPACTED CLAY LINERS

Introduction

Compacted soil liners are composed of clayey materials that are placed and compacted in layers called *lifts*. The materials used to construct soil liners include natural mineral materials (natural soils), bentonite-soil blends, and other material.

The most common type of compacted soil liner is one that is constructed from naturally occurring soils that contain a significant quantity of clay. Soil liner materials are excavated from locations called *borrow pits*. These borrow areas are located either onsite or offsite. The soil in the borrow pit may be used directly without processing or may be processed to alter the water content, break down large pieces of material, or remove oversized particles. Sources of natural soil liner materials include lacustrine deposits, glacial tills, aeolian materials, deltaic deposits, residual soils, and other types of soil deposits. Weakly cemented or highly weathered rocks, e.g., mudstones and shales, can also be used for soil liner materials, provided they are processed properly.

If the soils found in the vicinity of a waste disposal facility are not sufficiently clayey to be suitable for direct use as a soil liner material, a common practice is to blend natural soils available on or near a site with bentonite.

Critical CQC and CQA Issues

The CQC and CQA processes for soil liners are intended to accomplish three objectives:

1. Ensure that soil liner materials are suitable.
2. Ensure that soil liner materials are properly placed and compacted.
3. Ensure that the completed liner is properly protected.

Some of these issues, such as protection of the liner from desiccation after completion, simply require application of common-sense procedures. Other issues, such preprocessing of materials, are potentially much more complicated because, depending on the material, many construction steps may be involved. Furthermore, tests alone will not adequately address many of the critical CQC and CQA issues — visual observations by qualified personnel, supplemented by intelligently selected tests, provide the best approach to ensure quality in the constructed soil liner.

Construction Requirements

The steps involved in constructing a compacted clay liner may be summarized as follows:

1. The subgrade on which the soil liner will be placed should be properly prepared.
2. The materials employed in constructing the soil liner should be suitable and should conform to the plans and specifications for the project.
3. The soil liner material should be preprocessed, if necessary, to adjust the water content, to remove oversized particles, to break down clods of soil, or to add amendments such as bentonite.
4. The soil should be placed in lifts of appropriate thickness and then be properly remolded and compacted.
5. The completed soil liner should be protected from damage caused by desiccation or freezing temperatures.
6. The final surface of the soil liner should be properly prepared to support the next layer that will be placed on top of the soil liner.

Test Pads

Test pads are sometimes constructed and tested prior to construction of the full-scale compacted clay liner. The objectives of a test pad should be as follows:

1. To verify that the materials and methods of construction will produce a compacted soil liner that meets the hydraulic conductivity objectives defined for a project.

2. To verify that the proposed CQC and CQA procedures will result in a high-quality soil liner that will meet performance objectives.
3. To provide a basis of comparison for full-scale CQA: if the test pad meets the performance objectives for the liner (as verified by appropriate hydraulic conductivity tests) and the full-scale liner is constructed to standards that equal or exceed those used in building the test pad, then assurance is provided that the full-scale liner will also meet performance objectives.
4. If appropriate, a test pad provides an opportunity for the facility owner or contractor to demonstrate that unconventional materials or construction techniques will lead to a soil liner that meets performance objectives.

In terms of CQA, the test pad can provide an extremely powerful tool to ensure that performance objectives are met. The authors recommend a test pad for any project in which failure of the soil liner to meet performance objectives would have a potentially important, negative environmental impact.

Inspection of Borrow Sources Prior to Excavation

In order to determine the properties of the borrow soil, samples are often obtained from the potential borrow area for laboratory analysis prior to actual excavation but as part of the construction contract. Samples of soil should be taken for laboratory testing to ensure conformance with specifications.

The material tests that are normally performed on borrow soil are water content, Atterberg limits, particle size distribution, compaction curve, and hydraulic conductivity (Table 13.1). These tests are discussed in more detail in the succeeding subsections.

Water Content

It is important to know the water content of the borrow soils so that the need for wetting or drying the soil prior to compaction can be identified. The water content of the borrow soil is normally measured following the test methods mentioned in Table 13.1.

Atterberg Limits

Construction specifications for compacted soil liners often require a minimum value for the liquid limit and/or plasticity index of the soil. These limits are useful indicators of the amount of clay in the soil. Generally speaking, the greater the clay content, and the greater the activity of the clay component, the higher the Atterberg limits. These parameters are measured in the laboratory with the procedures outlined in ASTM D-4318.*

---

* American Society for Testing and Materials (ASTM) Standards are published in *ASTM Standards and Other Specifications and Test Methods on the Quality Assurance of Landfill Liner Systems*, ASTM, Philadelphia, PA, 527 pp.

Table 13.1 Materials Tests for Compacted Clay Liners

Parameter	ASTM test method	Title of ASTM test
Water content	D-2216	Laboratory determination of water (moisture) content of soil and rock
	D-4623	Determination of water (moisture) content of soil by the microwave oven method
	D-4944	Field determination of water (moisture) content of soil by the calcium carbide gas pressure tester method
	D-4959	Determination of water (moisture) content by direct heating method
Liquid limit, plastic limit, and plasticity index	D-4318	Liquid limit, plastic limit, and plasticity index of soils
Particle size distribution	D-422	Particle size analysis of soil
Compaction curve	D-698	Moisture-density relations for soils and soil-aggregate mixtures using 5.5 lb (2.48 kg) rammer and 12 in. (305 mm) drop
	D-1557	Moisture-density relations for soils and soil-aggregate mixtures using 10 lb (4.54 kg) rammer and 18 in. (457 mm) drop
Hydraulic conductivity	D-5084	Measurement of hydraulic conductivity of saturated porous materials using a flexible wall permeameter

From Daniel, D.E. and Koerner, R.M., *Waste Containment Facilities, Guidance for Construction, Quality Assurance, and Quality Control of Liner and Cover Systems*, American Society of Civil Engineers, New York, 1995. With permission.

Particle Size Distribution

Construction specifications for soil liners often place limits on the minimum percentage of fines, the maximum percentage of gravel, and in some cases the minimum percentage of clay. Particle size analysis is performed following the procedures in ASTM D-422.

Compaction Curve

Compaction curves, which are used to gauge the completeness of soil compaction, are developed utilizing the method of laboratory compaction testing required in the construction specifications. Standard compaction (ASTM D-698) and modified compaction (ASTM D-1557) are two common methods of laboratory compaction specified for soil liners.

Hydraulic Conductivity

The hydraulic conductivity of compacted samples of borrow material may be measured periodically to verify that the soil liner material can be compacted to achieve the required low hydraulic conductivity. Daniel and Koerner[1] describe sample preparation procedures.

Table 13.2   Recommended Minimum Testing Frequencies for Investigation of Clay Liner Borrow Source

Parameter	Frequency
Water content	1 Test per 2000 m³ or each change in material type
Atterberg limits	1 Test per 5000 m³ or each change in material type
Percentage fines	1 Test per 5000 m³ or each change in material type
Percent gravel	1 Test per 5000 m³ or each change in material type
Compaction curve	1 Test per 5000 m³ or each change in material type
Hydraulic conductivity	1 Test per 10,000 m³ or each change in material type

Note: 1 yd³ = 0.76 m³.

From Daniel, D.E. and Koerner, R.M., *Waste Containment Facilities, Guidance for Construction, Quality Assurance, and Quality Control of Liner and Cover Systems*, American Society of Civil Engineers, New York, 1995. With permission.

Testing Frequency

The CQA plan should stipulate the frequency of testing. Recommended minimum values are shown in Table 13.2. The tests listed in Table 13.2 are normally performed prior to construction as part of the characterization of the borrow source. However, if time or circumstances do not permit characterization of the borrow source prior to construction, the samples for testing are obtained during excavation or delivery of the soil materials.

Inspection during Excavation of Borrow Soil

It is strongly recommended that a qualified inspector who reports directly to the CQA engineer observe all excavation of borrow soil in the borrow pit. Often the best way to determine whether deleterious material is present in the borrow soil is to observe the excavation of the soil directly.

Preprocessing of Materials

Some soil liner materials are ready to be used for final construction immediately after they are excavated from the borrow pit. However, most materials require some degree of processing prior to placement and compaction of the soil.

Water Content Adjustment

Soils that are too wet must first be dried. If the water content needs to be reduced by no more than about three percentage points, the soil can be dried after it has been spread in a loose lift just prior to compaction. If the water content must be reduced by more than about 3 percentage points, it is recommended that drying take place in a separate processing area. The CQA inspectors should check to be sure that the soil is periodically mixed with a disc or rototiller to ensure uniform drying. The soil cannot be considered to be ready for placement and compaction unless the water is uniformly distributed; water content measurements alone do not ensure that water is uniformly distributed within the soil.

If the soil must be moistened prior to compaction, the same principles discussed above for drying apply; water content adjustment in a separate preprocessing area is recommended if the water content must be increased by more than about 3 percentage points. Inspectors should be careful to verify that water is distributed uniformly to the soil (a spreader bar on the back of a water truck is the recommended device for moistening soil uniformly), that the soil is periodically mixed with a disc or rototiller, and that adequate time has been allowed for uniform hydration of the soil.

Removal of Oversize Particles

Oversized stones and rocks should be removed from the soil liner material. Stones and rocks interfere with compaction of the soil and may create undesirable pathways for fluid to flow through the soil liner. The construction specifications should stipulate the maximum allowable size of particles in the soil liner material.

Oversized particles can be removed with mechanical equipment (e.g., large screens) or by hand. Inspectors should examine the loose lift of soil after the contractor has removed oversized particles to verify that oversized particles are not present.

Bentonite

Bentonite is a common additive to soil liner materials that do not contain enough clay to achieve the desired low hydraulic conductivity. Inspectors must ensure that the bentonite being used for a project is in conformance with specifications (i.e., is of the proper quality and gradation) and that the bentonite is uniformly mixed with soil in the required amounts.

The quality of bentonite is usually measured with some type of measurement of water adsorption ability of the clay, e.g., plate water adsorption test (ASTM E-946), liquid and plastic limits via ASTM D-4318, or free swell test.

The bentonite must usually also meet gradational requirements. Finely ground, powdered bentonite will behave differently when blended into soil than more coarsely ground, granular bentonite. Sieve tests on the raw bentonite received at a job site are recommended to verify gradation of the bentonite.

Mixing Bentonite

The best way to mix bentonite with soil is with a pugmill. A pugmill is a device for mixing dry materials, e.g., a concrete batch plant. A conveyor belt feeds soil into a mixing unit, and bentonite drops downward into the mixing unit. The materials are mixed in a large box that contains rotating rods with mixing paddles. Water may be added to the mixture in the pugmill, as well.

Measuring Bentonite Content

The best way to control the amount of bentonite mixed with soil is to measure the relative weights of soil and bentonite blended together at the time of mixing. After bentonite has been mixed with soil there are several techniques available to estimate the amount of bentonite in the soil. None of the techniques are particularly easy to use in all situations.

The recommended technique for measuring the amount of bentonite in soil is the methylene blue test.[2] The methylene blue test is a type of titration test. Methylene blue is slowly titrated into a material and the amount of methylene blue required to saturate the material is determined. The more bentonite in the soil, the greater the amount of methylene blue that must be added to achieve saturation. A calibration curve is developed between the amount of methylene blue needed to saturate the material and the bentonite content of the soil. The methylene blue test works very well when bentonite is added into a non-clayey soil. However, the amount of methylene blue that must be added to the soil is a function of the amount of clay present in the soil. If clay minerals other than bentonite are present, the clay minerals interfere with the determination of the bentonite content.

Measurement of hydraulic conductivity provides a means for verifying that enough bentonite has been added to the soil to achieve the desired low hydraulic conductivity. If insufficient bentonite has been added, the hydraulic conductivity should be unacceptably large.

Placement of Loose Lift of Soil

After a soil has been fully processed, the soil is hauled to the final placement area. Soil should not be placed in adverse weather conditions, e.g., heavy rain. Inspectors are usually responsible for documenting weather conditions during all earthwork operations. The surface on which the soil will be placed must be properly prepared and the material must be inspected after placement to make sure that the material is suitable. Then the CQA inspectors must also verify that the lift is not too thick.

Prior to placement of a new lift of soil, the surface of the previously compacted lift of soil liner should be roughened to promote good contact between the new and old lifts. Inspectors should observe the condition of the surface of the previously compacted lift to make sure that the surface has been scarified as required in the construction specifications. When soil is scarified, it is usually roughened to a depth of about 25 mm.

Material Tests and Visual Inspection

Material Tests

After a loose lift of soil has been placed, samples are periodically taken to confirm the properties of the soil liner material. These samples are in addition to

Table 13.3   Recommended Materials Tests for Soil Liner Materials Sampled after Placement in a Loose Lift (Just Before Compaction)

Parameter	Test method	Minimum testing frequency
Percent fines (Note 1)	ASTM D-1140	1 per 800 m³ (Notes 2 and 5)
Percent gravel (Note 3)	ASTM D-422	1 per 800 m³ (Notes 2 and 5)
Liquid and plastic limits	ASTM D-4318	1 per 800 m³ (Notes 2 and 5)
Percent bentonite (Note 4)	Alther²	1 per 800 m³ (Notes 2 and 5)
Compaction curve	As specified	1 per 4,000 m³ (Note 5)
Construction oversight	Observation	Continuous

Notes:

1. Percent fines is defined as percent passing the No. 200 sieve.
2. In addition, at least one test should be performed each day that soil is placed, and additional tests should be performed on any suspect material observed by CQA personnel.
3. Percent gravel is defined as percent retained on the No. 4 sieve.
4. This test is only applicable to soil-bentonite liners.
5. 1 yd³ = 0.76 m³.

From Daniel, D.E. and Koerner, R.M., *Waste Containment Facilities, Guidance for Construction, Quality Assurance, and Quality Control of Liner and Cover Systems*, American Society of Civil Engineers, New York, 1995. With permission.

samples taken from the borrow area (Table 13.2). The types of tests and frequency of testing are normally specified in the CQA documents. Table 13.3 summarizes recommended minimum tests and testing frequencies.

Visual Observations

Inspectors should position themselves near the working face of soil liner material as it is being placed. Inspectors should look for deleterious materials such as stones, debris, and organic matter. Continuous inspection of the placement of soil liner material is recommended to ensure that the soil liner material is of the proper consistency.

Corrective Action

If it is determined that the materials in an area do not conform with specifications, the first step is to define the extent of the area requiring repair. A sound procedure is to require the contractor to repair the lift of soil out to the limits defined by passing CQC/CQA tests. To define the limits of the area that requires repair, additional tests are often needed.

The usual corrective action is to wet or dry the loose lift of soil in place if the water content is incorrect. The water must be added uniformly. If the soil contains oversized material, oversized particles are removed from the material. If the soil lacks adequate plasticity, contains too few fines, contains too much gravel, or lacks adequate bentonite, the material is normally excavated and replaced.

Placement and Control of Loose Lift Thickness

Construction specifications normally place limits on the maximum thickness of a loose lift of soil, e.g., 225 mm. The best technique to verify that loose lifts are not too thick is for an inspector standing near the working face of soil being placed to observe the thickness of the lift.

Compaction of Soil

Compaction Equipment

The important parameters concerning compaction equipment are the type and weight of the compactor, the characteristics of any feet on the drum, and the weight of the roller per unit length of drummed surface. Inspectors should be particularly cognizant of the weight of compactor and length of feet on drummed rollers. Heavy compactors with long feet that fully penetrate a loose lift of soil are generally thought to be the best type of compactor to use for most clay liners.

Number of Passes

The compactive effort delivered by a roller is a function of the number of passes of the roller over a given area of soil. The number of passes of a compactor over the soil can have an important influence on the overall hydraulic conductivity of the soil liner. It is recommended that periodic observations be made of the number of passes of the roller over a given point. Approximately 3 observations per hectare per lift (one observation per acre per lift) is the recommended frequency of measurement.

Water Content and Dry Unit Weight

One of the most important CQA tests is measurement of water content and dry unit weight. Recommended minimum testing frequencies are listed in Table 13.4.

Some methods of measurement may introduce a systematic error. For example, the nuclear device for measuring water content may consistently produce a water content measurement that is too high if there is an extraneous source of hydrogen atoms besides water in the soil. It is important that devices that may introduce a significant systematic error be periodically correlated with measurements that do not have such error. Therefore, as indicated in Table 13.4, it is recommended that if the nuclear method or any of the rapid methods of water content measurement (Table 13.1) are used to measure water content, periodic correlation tests should be made with conventional overnight oven drying (ASTM D-2216).

**Table 13.4   Recommended Tests and Observations on Compacted Clay
Liners to Verify Proper Compaction of the Soil**

Parameter	Test method	Minimum testing frequency
Water content (rapid) (Note 1)	ASTM D-3017 ASTM D-4643 ASTM D-4944 ASTM D-4959	13/ha/lift (5/acre/lift) (Notes 2 and 7)
Water content (Note 3)	ASTM D-2216	One in every 10 rapid water content tests (Notes 3 and 7)
Total density (rapid) (Note 4)	ASTM D-2922 ASTM D-2937	13/ha/lift (5/acre/lift) (Notes 2, 4, and 7)
Total density (Note 5)	ASTM D-1556 ASTM D-1587 ASTM D-2167	One in every 20 rapid density tests (Notes 5, 6, and 7)
Number of passes	Observation	3/ha/lift (1/acre/lift) (Notes 2 and 7)
Construction oversight	Observation	Continuous

Notes:

1. ASTM D-3017 is a nuclear method, ASTM D-4643 is microwave oven drying, ASTM D-4944 is a calcium carbide gas pressure tester method, and ASTM D-4959 is a direct heating method. Direct water content determination (ASTM D-2216) is the standard against which nuclear, microwave, or other methods of measurements are calibrated for onsite soils.
2. In addition, at least one test should be performed each day soil is compacted and additional tests should be performed in areas for which CQA personnel have reason to suspect inadequate compaction.
3. Every tenth sample tested with ASTM D-3017, D-4643, D-4944, or D-4959 should be also tested by direct oven drying (ASTM D-2216) to aid in identifying any significant, systematic calibration errors.
4. ASTM D-2922 is a nuclear method and ASTM D-2937 is the drive cylinder method. These methods, if used, should be calibrated against the sand cone (ASTM D-1556) or rubber balloon (ASTM D-2167) for onsite soils. Alternatively, the sand cone or rubber balloon method can be used directly.
5. Every 20th sample tested with D-2922 or D-2937 should also be tested (as close as possible to the same test location) with the sand cone (ASTM D-1556) or rubber balloon (ASTM D-2167) to aid in identifying any systematic calibration errors with D-2922.
6. ASTM D-1587 is the method for obtaining an undisturbed sample. The section of undisturbed sample can be cut or trimmed from the sampling tube to determine bulk density. This method should not be used for soils containing any particles > 1/6th the diameter of the sample.
7. 1 acre = 0.4 ha.

From Daniel, D.E. and Koerner, R.M., *Waste Containment Facilities, Guidance for Construction, Quality Assurance, and Quality Control of Liner and Cover Systems*, American Society of Civil Engineers, New York, 1995. With permission.

Some methods of unit weight measurement may also introduce bias. It is recommended that unit weight be measured independently on occasion to provide a check against systematic errors (Table 13.4).

Hydraulic Conductivity Tests on Undisturbed Samples

Hydraulic conductivity tests per ASTM D-5084 are often performed on "undisturbed" samples of soil obtained from a single lift of compacted soil liner.

Test specimens are trimmed from the samples and are permeated in the laboratory. Compliance with the stated hydraulic conductivity criterion is checked.

This type of test is given far too much weight in most QA programs. Low hydraulic conductivity of samples taken from the liner is necessary for a well-constructed liner but is not sufficient to demonstrate that the large-scale field hydraulic conductivity is adequately low. For example, Elsbury et al.[3] performed laboratory hydraulic conductivity tests on undisturbed samples of a poorly constructed liner and found an average hydraulic conductivity of 1×10^{-9} cm/s, and yet the actual in-field value was 1×10^{-5} cm/s. The cause for the discrepancy was the existence of macro-scale flow paths in the field that were not simulated in the small-sized (75 mm diameter) laboratory test specimens.

Not only does the flow pattern through a 75 mm diameter test specimen not necessarily reflect flow patterns on a larger field scale, but the process of obtaining a sample for testing inevitably disturbs the soil. Layers are distorted, and gross alterations occur if significant gravel is present in the soil. The process of pushing a sampling tube into the soil densifies the soil, which lowers its hydraulic conductivity. The harder and drier the soil, the greater the disturbance. As a result of these various factors, the large scale, field hydraulic conductivity is almost always greater than or equal to the small scale, laboratory-measured hydraulic conductivity. The difference between values from a small laboratory scale and a large field scale depends on the quality of construction — the better the quality of construction, the less the difference.

Laboratory hydraulic conductivity tests on undisturbed samples of compacted liner can be valuable in some situations. For instance, for soil–bentonite mixes, the laboratory test provides a check on whether enough bentonite has been added to the mix to achieve the desired hydraulic conductivity. For soil liners in which a test pad is not constructed, the laboratory tests provide some verification that appropriate materials have been used and compaction was reasonable (but hydraulic conductivity tests by themselves do not prove this fact).

Hydraulic conductivity tests are typically performed at a frequency of 3 tests/ha/lift (1 test/acre/lift) or, for very thick liners (≥ 1.2 m) per every other lift. This is the recommended frequency of testing, if hydraulic conductivity testing is required. The CQA plan should stipulate the frequency of testing.

Repair of Holes from Sampling and Testing

A number of tests, e.g., from nuclear density tests and sampling for hydraulic conductivity, require that a penetration be made into a lift of compacted soil. It is extremely important that all penetrations be repaired. Backfill may consist of the soil liner material itself, granular or pelletized bentonite, or a mixture of bentonite and soil liner material. The backfill material should be placed in the hole requiring repair with a loose lift thickness not exceeding about 50 mm. The loose lift of soil should be tamped several times with a steel rod or other suitable device that compacts the backfill and ensures no bridging of material that would leave large air pockets. Next, a new lift of backfill should be placed and compacted. The process is repeated until the hole has been filled.

It is suggested that approximately 20% of all the repairs be inspected and that the backfill procedures be documented for these inspections. It is recommended that the inspector of repair of holes not be the same person who backfilled the hole.

Protection of Compacted Soil

Desiccation

There are several ways to prevent compacted soil liner materials from desiccating. The soil may be smooth rolled with a steel drummed roller to produce a thin, dense skin of soil on the surface. This thin skin of very dense soil helps to minimize transfer of water into or out of the underlying material.

Perhaps the best preventive measure is to water the soil periodically. Care must be taken to deliver water uniformly to the soil and not to create zones of excessively wet soil. Adding water by hand is not recommended because water is not delivered uniformly to the soil. An alternative preventive measure is to cover the soil temporarily with a geomembrane, moist geotextile, or moist soil.

If soil has been desiccated to a depth less than or equal to the thickness of a single lift, the desiccated lift may be disked, moistened, and recompacted. However, disking may produce large, hard clods of clay that will require pulverization. If deeper desiccation has occurred, it may be necessary to remove and replace the damaged material.

Freezing Temperatures

Frozen soil should never be used to construct soil liners. Frozen soils form hard pieces that cannot be properly remolded and compacted. Soil may freeze after it has been compacted in a thawed state. Freezing of soil liner materials can produce significant increases in hydraulic conductivity. Soil liners must be protected from freezing before and after construction. If superficial freezing takes place on the surface of a lift of soil, the surface may be scarified and recompacted. If an entire lift has been frozen, the entire lift should be disked, pulverized, and recompacted. If the soil is frozen to a depth greater than one lift, it may be necessary to strip away and replace the frozen material.

Final Approval

Upon completion of the soil liner, the soil liner should be accepted and approved by the CQA engineer prior to deployment or construction of the next overlying layer.

GEOMEMBRANE LINERS

This section focuses upon the manufacturing quality assurance (MQA) aspects of geomembrane formulation, manufacture and fabrication, and on the construc-

Table 13.5   Types of Commonly Used Geomembranes and Their Approximate Weight Percentage Formulations[1]

Geomembrane type	Resin	Plasticizer	Filler	Carbon black or pigment	Additives
HDPE	95–98	0	0	2–3	0.25–1.0
VLDPE	94–96	0	0	2–3	1–4
Other extruded types[2]	95–98	0	0	2–3	1–2
PVC	50–70	25–35	0–10	2–5	2–5
CSPE[3]	40–60	0	40–50	5–40	5–15
Other calendered types[2]	40–97	0–30	0–50	2–30	0–7

[1] Note that this table should not be directly used for MQA or CQA Documents, since neither the Agency nor the authors intend to provide prescriptive formulations for manufacturers and their respective geomembranes.
[2] Other geomembranes than those listed in this table will be described in the appropriate section.
[3] CSPE geomembranes are generally fabric (scrim) reinforced.

From Daniel, D.E. and Koerner, R.M., *Waste Containment Facilities, Guidance for Construction, Quality Assurance, and Quality Control of Liner and Cover Systems*, American Society of Civil Engineers, New York, 1995. With permission.

tion quality assurance (CQA) of the complete installation of the geomembranes in the field. Note that in the literature geomembranes are also called *flexible membrane liners* (FMLs), but the more generic name of geomembranes will be used throughout this chapter.

Types of Geomembranes and Their Formulations

All geomembranes are actually formulations of a parent resin (from which they derive their generic name) and several other ingredients. The most commonly used geomembranes for solid and liquid waste containment are listed in Table 13.5 according to their commonly referenced acronyms. It must be recognized that Table 13.5 and the references to it in the text to follow are meant to reflect on the current state-of-the-art. The values mentioned are not meant to be prescriptive and future research and development may result in substantial changes.

Daniel and Koerner[1] provide details on MQC/MQA recommendations for all the major geomembrane types. For brevity, and to illustrate the process just one of the geomembranes (high density polyethylene, HDPE) is discussed in this chapter.

As noted in Table 13.5, high density polyethylene (HDPE) geomembranes are made from polyethylene resin, carbon black, and additives.

Resin

The polyethylene resin used for HDPE geomembranes is prepared by low pressure polymerization of ethylene as the principal monomer and having the characteristics listed in ASTM D-1248. The resin is usually supplied to the manufacturer or formulator in an opaque pellet form.

Regarding the preparation of a specification or MQA document for the resin component of an HDPE geomembrane, the following items should be considered:

1. The polyethylene resin, which is covered in ASTM D-1248, is to be made from virgin, uncontaminated ingredients.
2. The quality control tests performed on the incoming resin will typically be density (ASTM D-1505) and melt flow index which is ASTM D-1238.
3. Typical natural densities of the various resins used are between 0.934 and 0.940 g/cc. Note that according to ASTM D-1248 this is Type II polyethylene and is classified as medium density polyethylene.
4. Typical melt flow index values are between 0.1 and 1.0 g/10 min as per ASTM D-1238, Cond. 190/2.16.
5. Other tests which can be considered for quality control of the resin are melt flow ratio (comparing high-to-low weight melt flow values), notched constant tensile load test as per ASTM D-5397, and a single point notched constant load test — see Hsuan and Koerner[4] for details. The latter tests would require a plaque to be made from the resin from which test specimens are taken. The single point notched constant load test is then performed at 30% yield strength and the test specimens are currently recommended not to fail within 200 hours.
6. Additional quality control certification procedures by the manufacturer (if any) should be implemented and followed.
7. The frequency of performing each of the preceding tests should be covered in the MQC plan and it should be implemented and followed.
8. An HDPE geomembrane formulation should consist of at least 97% of polyethylene resin. As seen in Table 13.5 the balance is carbon black and additives. No fillers, extenders, or other materials should be mixed into the formulation.
9. It should be noted that by adding carbon black and additives to the resin, the density of the final formulation is generally 0.941 to 0.954 g/cc. Since this numeric value is now in the high density polyethylene category according to ASTM D-1248, geomembranes of this type are commonly referred to as high density polyethylene (HDPE).
10. Regrind or rework chips (which have been previously processed by the same manufacturer but never used as a geomembrane, or other) are often added to the extruder during processing.
11. Reclaimed material (which is post-consumer polymer material that has seen previous service life and is recycled) should never be allowed in the formulation in any quantity.

Carbon Black

Carbon black is added into an HDPE geomembrane formulation for general stabilization purposes, particularly for ultraviolet light stabilization. It is sometimes added in a powder form at the geomembrane manufacturing facility during processing, or (generally) it is added as a preformulated concentrate in pellet form. The latter is the usual case.

Regarding the preparation of a specification or MQA document for the carbon black component of HDPE geomembranes, the following items should be considered.

1. The carbon black used in HDPE geomembranes should be a Group 3 category, or lower, as defined in ASTM D-1765.
2. Typical amounts of carbon black are from 2.0 to 3.0% by weight per ASTM D-1603. Values less than 2.0% do not appear to give adequate long-term ultraviolet protection; values greater than 3.0% begin to adversely effect physical and mechanical properties.
3. Carbon black dispersion requirements in the final HDPE geomembrane are evaluated according to ASTM D-5596.
4. In the event that the carbon black is mixed into the formulation in the form of a concentrate rather than a powder, the carrier resin of the concentrate should be the same generic type as the base polyethylene resin.

Additives

Additives are introduced into an HDPE geomembrane formulation for the purposes of oxidation prevention, long-term durability, and as a lubricant and/or processing aid during manufacturing. It is quite difficult to write a specification for HDPE geomembranes around a particular additive, or group of additives, because they are generally proprietary. Furthermore, there is research and development ongoing in this area and thus additives are subject to change over time.

If additives are included in a specification or MQA document, the description must be very general as to the type and amount. However, the amount can probably be bracketed as to an upper value.

1. The nature of the additive package used in the HDPE compound may be requested of the manufacturer.
2. The maximum amount of additives in a particular formulation should not exceed 1.0% by weight.

Manufacturing

Once the specific type of geomembrane formulation that is specified has been thoroughly mixed, it is then manufactured into a continuous sheet. The two major processes used for manufacturing of the various types of sheets of geomembranes are variations of either extrusion (e.g., for HDPE, VLDPE, and LLDPE) or calendering (e.g., for PVC, CSPE, and PP). The actual manufacturing and MQC requirements for the manufacturing phase are product-specific and are described in detail by Daniel and Koerner.[1]

Shipment, Handling, and Site Storage

The geomembrane rolls or pallets are shipped to the job site, offloaded, and temporarily stored at a remote location on the job site. Regarding items for a specification or CQA document, the following applies:

1. Unloading of rolls or pallets at the job site's temporary storage location should be such that no damage to the geomembrane occurs.

2. Pushing, sliding, or dragging of rolls or pallets of geomembranes should not be permitted.
3. Offloading at the job site should be performed with cranes or forklifts in a workmanlike manner such that damage does not occur to any part of the geomembrane.
4. Temporary storage at the job site should be in an area where standing water cannot accumulate at any time.
5. The ground surface should be suitably prepared such that no stones or other rough objects that could damage the geomembranes are present.
6. Temporary storage of rolls of HDPE or VLDPE geomembranes in the field should not be so high that crushing of the core or flattening of the rolls occur. This limit is typically 5 rolls high.
7. Temporary storage of pallets of PVC or CSPE-R geomembranes by stacking should not be permitted.
8. Suitable means of securing the rolls or pallets should be used such that shifting, abrasion or other adverse movement does not occur.
9. If storage of rolls or pallets of geomembranes at the job site is longer than 6 months, a sacrificial covering or temporary shelter should be provided for protection against precipitation, ultraviolet exposure, and accidental damage.

Acceptance and Conformance Testing

It is the primary duty of the installation contractor, via the CQC personnel, to see that the geomembrane supplied to the job site is the proper material that was called for in the contract, as specified by the plans and specifications. It is also the duty of the CQA engineer to verify this material to be appropriate. Clear marking should identify all rolls or pallets with all required information. A complete list of roll numbers should be prepared for each material type.

Upon delivery of the rolls or pallets of geomembrane, the CQA Engineer should ensure that conformance test samples are obtained and sent to the proper laboratory for testing. This will generally be the laboratory of the CQA firm, but may be that of the CQC firm if so designated in the CQA documents. Alternatively, conformance testing could be performed at the manufacturer's facility and when completed, the particular lot should be marked for the particular site under investigation.

The following items should be considered for a specification or CQA document with regard to acceptance and conformance testing.

1. The particular tests selected for acceptance and conformance testing can be all of those listed previously, but this is rarely the case since MQC and MQA testing should have preceded the field operations. However, at a minimum, the following tests are recommended for field acceptance and conformance testing for the particular geomembrane type.
   a. HDPE and VLDPE: thickness (ASTM D-5199), tensile strength and elongation (ASTM D-638) and possibly puncture[5] and tear resistance (ASTM D-1004, Die C).
   b. PVC: thickness (ASTM D-5199), tensile strength and elongation (ASTM D-882), tear resistance (ASTM D-1004, Die C).

    c. CSPE-R: thickness (ASTM D-5199), tensile strength and elongation (ASTM D-751), ply adhesion (ASTM D-413, Machine Method, Type A).

2. The method of geomembrane sampling should be prescribed. For geomembranes on rolls, 1 m (3 ft) from the entire width of the roll on the outermost wrap is usually cut and removed. For geomembranes folded on pallets, the protective covering must be removed, the uppermost accordion-folded section opened and an appropriate size sample taken. Alternatively, factory seam retains can be shipped on top of fabricated panels for easy access and use in conformance testing.

3. The machine direction must be indicated with an arrow on all samples using a clearly visible permanent marker.

4. Samples are usually taken on the basis of a stipulated area of geomembrane, e.g., one sample per 10,000 m². Alternatively, one could take samples at the rate of one per lot, however, a lot must be clearly defined. One possible definition could be that a lot is a group of consecutively numbered rolls or panels from the same manufacturing line.

5. All conformance test results should be reviewed, accepted, and reported by the CQA Engineer before deployment of the geomembrane.

6. Any nonconformance of test results should be reported to the Owner/Operator. The method of a resolution of such differences should be clearly stated in the CQA document. One possible guidance document for failing conformance tests could be ASTM D-4759.

Placement

When the subgrade or subbase (either soil or some other geosynthetic) is approved as being acceptable, the rolls or pallets of the temporarily stored geomembranes are brought to their intended location, unrolled, or unfolded, and accurately spotted for field seaming.

Subgrade (Subbase) Conditions

Before beginning to move the geomembrane rolls or pallets from their temporary storage location at the job site, the soil subgrade (or other subbase material) should be checked for its preparedness. Some items recommended for a specification or CQA document include the following:

1. The soil subgrade shall be of the specified grading, moisture content and density as required by the installer and as approved by the CQA engineer for placement of the geomembrane. Attention should be given to isolated stones or gravel particles on the surface or protruding above the surface. These should be no larger than 12 mm (0.5 in.) for HDPE and CSPE-R. They can be somewhat larger for VLDPE and PVC.

2. Construction equipment deploying the rolls or pallets shall not deform or rut the soil subgrade excessively. Tire or track deformations beneath the geomembrane should not be greater than 25 mm in depth.

3. The geomembrane shall not be deployed on frozen subgrade where ruts are greater than 12 mm in depth.

4. When placing the geomembrane on another geosynthetic material, construction equipment should not be permitted to ride directly on the lower geosynthetic material. In cases where rolls must be moved over previously placed geosynthetics, it is necessary to move materials by hand or by using small pneumatic-tired lifting units. Tire inflation pressures should be limited to a maximum value of 40 kPa.
5. Underlying geosynthetic materials (such as geotextiles or geonets) should have all folds, wrinkles, and other undulations removed before placement of the geomembrane.
6. Care and planning should be taken to unroll or unfold the geomembrane close to its intended, and final, position.

Geomembranes expand when they are heated and contract when they are cooled. This expansion and contraction must be considered when placing, seaming, and backfilling geomembranes in the field.

Seaming and Joining

The field seaming of the deployed geomembrane rolls or panels is a critical aspect of their successful functioning as a barrier to liquid (and sometimes vapor) flow. This section describes the various seaming methods in current use, references a recently published EPA Technical Guidance Document on seam fabrication techniques,[5] and describes the concept and importance of test strips (or trial seams).

The fundamental mechanism of seaming polymeric geomembrane sheets together is to temporarily reorganize, i.e., melt, the polymer structure of the two surfaces to be joined in a controlled manner that, after the application of pressure and after the passage of a certain amount of time, results in the two sheets being bonded together. The methods of seaming are given in Table 13.6.

There are four general categories of seaming methods: *extrusion welding*, *thermal fusion or melt bonding*, *chemical fusion*, and *adhesive seaming*.

Table 13.6   Fundamental Methods Of Joining Polymeric Geomembranes

Thermal processes	Chemical processes
Extrusion	Chemical
• Fillet	• Chemical fusion
• Flat	• Bodied chemical fusion
Fusion	Adhesive
• Hot wedge	• Chemical adhesive
• Hot air	• Contact adhesive

From Daniel, D.E. and Koerner, R.M., *Waste Containment Facilities, Guidance for Construction, Quality Assurance, and Quality Control of Liner and Cover Systems*, American Society of Civil Engineers, New York, 1995. With permission.

Extrusion fillet welding is presently used exclusively on geomembranes made from polyethylene. A ribbon of molten polymer is extruded over the edge of the two surfaces to be joined. The molten extrudate causes the surfaces of the sheets to become hot and melt, after which the entire mass cools and bonds together. It should be noted that extrusion fillet seaming is essentially the only practical method for seaming polyethylene geomembrane patches, for seaming in poorly accessible areas such as sump bottoms and around pipes and for seaming of extremely short seam lengths. Temperature and seaming rate both play important roles in obtaining an acceptable bond; excessive melting weakens the geomembrane and inadequate melting results in poor extrudate flow across the seam interface and low seam strength. The polymer used for the extrudate is also very important and should be the same polyethylene compound used to make the geomembrane. The designer should specify acceptable extrusion compounds and how to evaluate them in the specifications and CQA documents.

There are two thermal fusion or melt-bonding methods that can be used on all thermoplastic geomembranes. In both of them, portions of the opposing surfaces are truly melted. This being the case, temperature, pressure, and seaming rate all play important roles in that excessive melting weakens the geomembrane and inadequate melting results in low seam strength. The *hot wedge*, or hot shoe, method consists of an electrically heated resistance element in the shape of a wedge that travels between the two sheets to be seamed. As it melts the surface of the two sheets being seamed, a shear flow occurs across the upper and lower surfaces of the wedge. Roller pressure is applied as the two sheets converge at the tip of the wedge to form the final seam. Hot wedge units are controllable as far as temperature, amount of pressure applied, and travel rate. A standard hot wedge creates a single uniform width seam, while a dual hot wedge (or "split" wedge) forms two parallel seams with a uniform unbonded space between them. This space is then used to evaluate seam quality and continuity of the seam by pressurizing the unbonded space with air and monitoring any drop in pressure that may signify a leak in the seam.

The *hot air* method makes use of a device consisting of a resistance heater, a blower, and temperature controls to force hot air between two sheets to melt the opposing surfaces. Immediately following the melting of the surfaces, pressure is applied to the seamed area to bond the two sheets. As with the hot wedge method, both single and dual seams can be produced. In selected situations, this technique may also be used to temporarily "tack" weld two sheets together until the final seam or weld is made and accepted.

Regarding the *chemical fusion* seam types; chemical fusion seams make use of a liquid chemical applied between the two geomembrane sheets to be joined. After a few seconds, required to soften the surface, pressure is applied to make complete contact and bond the sheets together. As with any of the chemical seaming processes to be described, the two adjacent materials to be bonded are transformed into a viscous phase. Care must be used to see that the proper amount of chemical is applied in order to achieve the desired results. *Bodied chemical*

Table 13.7 Possible Field Seaming Methods for Various Geomembranes

Type of seaming method	HDPE	VLDPE	Other PE	PVC	CSPE-R	Other flexible
Extrusion (fillet and flat)	A	A	A	n/a	n/a	A
Thermal fusion (hot wedge and hot air)	A	A	A	A	A	A
Chemical (chemical and bodied chemical)	n/a	n/a	n/a	A	A	A
Adhesive (chemical and contact)	n/a	n/a	n/a	A	A	A

Note: A = method is applicable; n/a = method is not applicable.

From Daniel, D.E. and Koerner, R.M., *Waste Containment Facilities, Guidance for Construction, Quality Assurance, and Quality Control of Liner and Cover Systems*, American Society of Civil Engineers, New York, 1995. With permission.

fusion seams are similar to chemical fusion seams except that 1% to 20% of the parent lining resin or compound is dissolved in the chemical and then is used to make the seam. The purpose of adding the resin or compound is to increase the viscosity of the liquid for slope work and/or adjust the evaporation rate of the chemical. This viscous liquid is applied between the two opposing surfaces to be bonded. After a few seconds, pressure is applied to make complete contact. *Chemical adhesive* seams make use of a dissolved bonding agent (an adherent) in the chemical or bodied chemical that is left after the seam has been completed and cured. The adherent thus becomes an additional element in the system. *Contact adhesives* are applied to both mating surfaces. After reaching the proper degree of tackiness, the two sheets are placed on top of one another, followed by application of roller pressure. The adhesive forms the bond and is an additional element in the system.

Other emerging seaming methods use ultrasonic, electrical conduction, and magnetic induction energy sources. Since these methods are in the developmental stage, they will not be described further in this document; see Reference 6 for further details.

In order to gain an overview as to which seaming methods are used for the various thermoplastic geomembranes described in this document, Table 13.7 is offered. It is generalized, but it is used to introduce the primary seaming methods versus the type of geomembrane that is customarily seamed by that method.

Test Strips and Trial Seams

Test strips and trial seams, also called qualifying seams, are considered to be an important aspect of CQC/CQA procedures. They are meant to serve as a prequalifying experience for personnel, equipment, and procedures for making seams on the identical geomembrane material under the same climatic conditions as the actual field production seams will be made. The test strips are usually made on two narrow pieces of excess geomembrane varying in length between

1 and 3 m. The test strips should be made in sufficient lengths, preferably as a single continuous seam, for all required testing purposes.

Destructive Test Methods for Seams

By destructively testing geomembrane seams it is meant to actually cut out (i.e., to sample) and remove a portion of the completed production seam, and then to further cut the sample into appropriately sized test specimens. These specimens are then tested according to a specified procedure to failure.

A possible procedure is to select the sampling location and cut two closely spaced 25 mm wide test specimens from the seam. The distance between these two test specimens is defined later. The individual specimens are then tested in a peel mode using a field tensiometer. If the results are acceptable, the complete seam between the two field test specimens is removed and properly identified and distributed. If either test specimen fails, two new locations on either side of the failed specimen(s) are selected until acceptable seams are located. The seam distance between acceptable seams is usually repaired by cap-stripping but other techniques are also possible. The exact procedure must be stipulated in the specifications or CQA document.

The hole created in the production seam from which the test sample was originally taken must be patched in an appropriate manner. Seams of such patches are themselves candidates for field sampling and testing. If this is done, one would have the end result of patch on a patch, which is a rather unsightly and undesirable condition.

By far the most commonly used sampling strategy is the "fixed increment sampling" method. In this method, a seam sample is taken at fixed increments along the total length of the seams. Increments usually range from 75 to 225 m with a commonly specified value being one destructive test sample every 150 m. Note that this value can be applied either directly to the record drawings during layout of the seams, to each seaming crew as they progress during the work period, or to each individual seaming device. Once the increment is decided upon, it can be held regardless of the location upon which it falls, or one can randomly select the sample location within the increment. Seam tests are usually "shear" and "peel" (Figure 13.2).

Shear Testing of Geomembrane Seams

Insofar as the shear testing of nonreinforced geomembrane seams (HDPE, VLDPE and PVC), all use a 25 mm wide test specimen with the seam being centrally located within the testing grips. The test specimen is tensioned, at its appropriate strain rate, until failure occurs. If the seam delaminates (i.e., pulls apart in a seam separation mode), the seam fails in what is called a "non-film tear bond," or non-FTB. In this case, it is rejected as a failed seam. Details on various types of seam failures and on the interpretation of FTB are found in Reference 7. Conversely, if the seam does not delaminate, but fails in the adjacent sheet

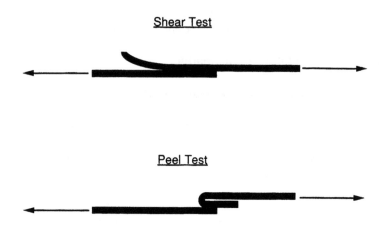

Figure 13.2   Shear and peel tests for geomembranes.

material on either side of the seam, it is an acceptable failure mode, i.e., called a "film tear bond," or FTB, and the seam strength is then calculated. The resulting value of seam shear strength is then compared to the required seam strength (which is the usual case) or to the strength of the unseamed geomembrane sheet.

The contract plans, specifications or CQA documents should give the minimum allowable seam shear strength or seam efficiency. In the latter case, the guidance listed below can be used whereby percentages of seam shear efficiencies (or values) are listed:

HDPE = 95% of specified minimum yield strength
VLDPE = typically 8,300 kPa
PVC = 80%
CSPE-R = 80% (for 3-ply reinforced)
EIA-R = 80%

Generally an additional requirement of a film tear bond, or FTB, will also be required in addition to a minimum strength value. This means that the failure must be located in the sheet material on either side of the seam and not within the seam itself. Thus the seam cannot delaminate.

Lastly, the number of failures allowed per number of tests conducted should be addressed. If sets of 5 test specimens are performed for each field sample, many specifications allow for one failure out of the five tested with the average of all 5 tests being equal, or greater, than the required value. If the failure number is larger, then the plans, specifications, or CQA documents must be clear on the implications.

Peel Testing of Geomembrane Seams

Peel testing of specimens taken from field fabricated geomembrane seams represent a quality control type of index test. Such tests are not meant to simulate *in situ* performance but are very important indicators of the overall quality of the seam. The recommended peel tests for HDPE, PVC, CSPE-R, and EIA-R seams, along with the unseamed sheet material in tension are given in Table 13.8. The VLDPE data was included in a way so as to parallel the HDPE testing protocol.

Insofar as the peel testing of geomembrane seams is concerned, it is seen that all of the geomembranes listed have a 25 mm width test specimen. Furthermore, the specimen lengths and strain rate are also equal for all geomembrane types. The only difference is that HDPE and VLDPE use the thickness of the geomembrane to calculate a tensile strength value in stress units, whereas PVC, CSPE-R and EIA-R calculate the tensile strength value in units of force per unit width, i.e., in units of pounds per linear inch of seam.

In a peel test the test specimen is tensioned, at its appropriate strain rate, until failure occurs. If the seam delaminates (i.e., pulls apart in a seam separation mode), it is called a "non-film tear bond," or non-FTB, and is recorded accordingly. Conversely, if the seam does not delaminate, but fails in the adjacent sheet material on either side of the seam it is called a "film tear bond" or FTB and the seam strength is calculated. Details on various types of seam failures and on the interpretation of FTB are found in Reference 7. The seam strength is the maximum force attained divided by the specimen width (resulting in units of force per unit width), or by the specimen cross sectional area (resulting in units of stress). The former procedure is the most common, i.e., peel strengths are measured in force per unit width units. The resulting value of seam peel strength is then compared to a specified value (the usual case) or to the strength of the unseamed geomembrane sheet. The testing procedures for obtaining these values are listed in Table 13.8.

The contract plans, specifications or CQA documents should give the minimum allowable seam peel strength efficiency. As a minimum, the guidance listed below can be used whereby percentage peel efficiencies (or values) are listed as follows:

HDPE = 62% of specified minimum yield strength and FTB
VLDPE = typically 6,900 kPa
PVC = 1.8 N/m (10 lb/in)
CSPE-R = 1.8 N/m (10 lb/in) or FTB
EIA-R = 1.8 N/m (10 lb/in)

Non-Destructive Seam Testing

A variety of non-destructive methods of seam testing are available.[1] Perhaps the most significant is pressure testing of the air gap left between a double hot wedge weld. A dual hot-wedge welding device forms two welds, about 10 to 15 mm apart, which leaves a 10 to 15 mm gap between. The gap is pressurized to a prescribed amount which depends on the type and thickness of the geomembrane. Seam lengths of approximately 100 to 300 m can be tested in this manner. If the seam holds the pressure, the seam passes. If not, further testing is performed to isolate the problem area, which is patched.

Table 13.8   Recommended Test Method Details for Geomembrane Seams in Shear and in Peel and for Unseamed Sheet

	Type of test				
	HDPE	VLDPE	PVC	CSPE-R	EIA-R
Shear test on seams					
ASTM test method	D4437	D4437	D3083	D751	D751
Specimen shape	Strip	Strip	Strip	Grab	Grab
Specimen width (in.)	1.00	1.00	1.00	4.00 (1.00 grab)	4.00 (1.00 grab)
Specimen length (in.)	6.00 + seam	6.00 + seam	6.00 + seam	9.00 + seam	9.00 + seam
Gage length (in.)	4.00 + seam	4.00 + seam	4.00 + seam	6.00 + seam	6.00 + seam
Strain rate (ipm)	2.0	20	20	12	12
Strength (psi) or (ppi)	Force/(1.00 × t)	Force/(1.00 × t)	Force/(1.00 × t)	Force	Force
Peel test on seams					
ASTM test method	D4437	D4437	D413	D413	D751
Specimen shape	Strip	Strip	Strip	Strip	Strip
Specimen width (in.)	1.00	1.00	1.00	1.00	1.00
Specimen length (in.)	4.00	4.00	4.00	4.00	4.00
Gage length (in.)	n/a	n/a	n/a	n/a	n/a
Strain rate (ipm)	2.0	20	2.0	2.0	2.0
Strength (psi) or (ppi)	Force/(1.00×t)	Force/(1.00 × t)	Force/1.00	Force/1.00	Force/1.00
Tensile test on sheet					
ASTM test method	D638	D638	D882	D751	D751
Specimen shape	Dumbbell	Dumbbell	Strip	Grab	Grab
Specimen width (in.)	0.25	0.25	1.00	4.00 (1.00 grab)	4.00 (1.00 grab)
Specimen length (in.)	4.50	4.50	6.00	6.00	6.00
Gage length (in.)	1.30	1.30	2.00	3.00	3.00
Strain rate (ipm)	2.0	20	20	12	12
Strength (psi) or (lb)	Force/(0.25 t)	Force/(0.25 t)	Force/(1.00 t)	Force	Force
Strain (in./in.)	Elong./1.30	Elong./1.30	Elong./2.00	Elong./3.00	Elong./3.00
Modulus (psi)	From graph	From graph	From graph	n/a	n/a

Note: n/a = not applicable; t = geomembrane thickness; psi = pounds/square inch of specimen cross section; ppi = pounds/linear inch width of specimen; ipm = inches/minute; Force = maximum force attained at specimen failure (yield or break).

From Daniel, D.E. and Koerner, R.M., *Waste Containment Facilities, Guidance for Construction, Quality Assurance, and Quality Control of Liner and Cover Systems*, American Society of Civil Engineers, New York, 1995. With permission.

Protection and Backfilling

The field deployed and seamed geomembrane must be backfilled with soil or covered with a subsequent layer of geosynthetics in a timely manner after its acceptance by the CQA personnel. If the covering layer is soil, it will generally be a drainage material like sand or gravel depending upon the required permeability of the overlying layer. Depending upon the particle size, hardness, and angularity of this soil, a geotextile or other type of protection layer may be necessary. If the covering layer is a geosynthetic, it will generally be a geonet or geocomposite drain, which is usually placed directly upon the geomembrane. This is obviously a critical step since geomembranes are relatively thin materials with puncture and tear strengths of finite proportions. Specifications should be very clear and unequivocal regarding this final step in the installation survivability of geomembranes.

CONCLUSIONS

Proper quality assurance and quality control includes manufacturing of geosynthetic materials and construction of both natural and synthetic materials. The level of detail that is considered good QA/QC practice (i.e., the current state of the practice) is very high. This paper has highlighted the recommended QA/QC procedures for just two of the many possible components of a modern, engineered municipal solid waste landfill.

ACKNOWLEDGMENTS

The work described in this chapter was supported by cooperative agreement CR-814456-01-0 by the U.S. Environmental Protection Agency (EPA), Risk Reduction Engineering Laboratory, Cincinnati, OH. Mr. David A. Carson was the project officer. Mr. Robert Landreth of the EPA provided important input to the work.

REFERENCES

1. Daniel, D. E. and Koerner, R. M. (1993), Technical Guidance Document: Quality Assurance and Quality Control for Waste Containment Facilities, U.S. Environmental Protection Agency, Washington, D.C., EPA/600/R-93/182, 305 p.
2. Alther, G. R. (1983), The Methylene Blue Test for Bentonite Liner Quality Control, *Geotechnical Testing Journal*, 6 (3), pp. 133–143.
3. Elsbury, B. R., Daniel, D. E., Sraders, G. A., and Anderson, D. C. (1990), Lessons Learned from Compacted Clay Liner, *Journal of Geotechnical Engineering*, 116 (11), pp. 1641–1660.
4. Hsuan, Y. and Koerner, R. M. (1992), Stress Cracking Potential and Behavior of HDPE Geomembranes, U.S. Environmental Protection Agency, Cincinnati, OH.

5. FTM Std. 101C, Puncture Resistance and Elongation Test, Federal Test Method 2065, American Society for Testing and Materials, Philadelphia, March 13, 1980.
6. U.S. Environmental Protection Agency (1991), Inspection Techniques for the Fabrication of Geomembrane Field Seams, EPA Technical Guidance Document, EPA/530/SW-91/051.
7. Haxo, H. E. (1988), Lining of Waste Containment and Other Impoundment Facilities, U.S. EPA/600/2-88/052, Washington, D.C.

CHAPTER 14

Cover Systems for Waste Management Facilities

Robert E. Landreth

INTRODUCTION

Proper closure is essential to complete a filled hazardous waste landfill or a municipal solid waste landfill. Research has established minimum requirements needed to meet the stringent, necessary, closure regulations in the U.S. In designing the landfill cover for hazardous waste landfills, the objective is to limit the infiltration of water to the waste to minimize creation of leachate that could possibly escape to groundwater sources.

Minimizing leachates in a closed waste management unit requires that liquids be kept out and that the leachate that does exist be detected, collected, and removed. Where the waste is above the groundwater zone, a properly designed and maintained cover can prevent (for practical purposes) water from entering the landfill and, thus, minimize the formation of leachate.

The cover system must be devised at the time the site is selected and the plan and design of the landfill containment structure is chosen. The location, the availability of soil with a low permeability or hydraulic conductivity, the stockpiling of good topsoil, the availability and use of geosynthetics to improve performance of the cover system, the height restrictions to provide stable slopes, and the use of the site after the post-closure care period are typical considerations. The goals of the cover system are to minimize further maintenance and to protect human health and the environment.

Subparts G, K, and N of the Resource Conservation and Recovery Act (RCRA) Subtitle C regulations form the basic requirements for cover systems being designed and constructed today. Comprehensive Environmental Response, Compensation, and Liability Act (CERCLA) regulations refer to the RCRA Subtitle C regulations but other criteria, primarily approved state requirements,

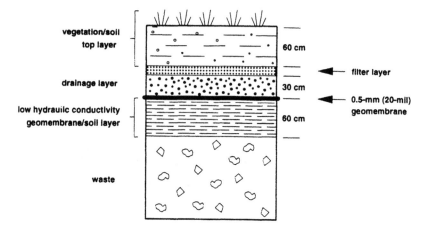

Figure 14.1   EPA recommended landfill cover design.

also have to be evaluated for applicability. The municipal solid waste landfill regulations (Subtitle D) base cover requirements on the leachate strategy and the hydraulic conductivity of the bottom liner. There is sufficient flexibility built into the regulations for innovative design.

RECOMMENDED DESIGN FOR SUBTITLE C FACILITIES

After the hazardous waste management unit is closed, the U.S. Environmental Protection Agency (EPA) recommends[1] that the final cover (Figure 14.1) consist of, from bottom to top:

1. *A Low Hydraulic Conductivity Geomembrane/Soil Layer.* A 60-cm (24-in.) layer of compacted natural or amended soils with a hydraulic conductivity of 1×10^{-7} cm/s in intimate contact with a minimum 0.5-mm (20-mil) geomembrane liner.
2. *A Drainage Layer.* A minimum 30-cm (12-in.) soil layer having a minimum hydraulic conductivity of 1×10^{-2} cm/s, or a layer of geosynthetic material having the same characteristics.
3. *A Top Vegetation/Soil Layer.* A top layer with vegetation (or an armored top surface) and a minimum of 60-cm (24-in.) of soil graded at a slope between 3 and 5%.

Because the design of the final cover must consider the site, the weather, the character of the waste, and other site-specific conditions, these minimum recommendations may be altered providing the alternative design is equivalent to the EPA-recommended design or will meet the intent of the regulations. EPA encourages design innovation and will accept an alternative design provided the owner or operator demonstrates the new design's equivalency. For example, in extremely arid regions, a gravel top surface might compensate for reduced vegetation, or the middle drainage layer might be expendable. Where burrowing animals might damage the geomembrane/low hydraulic conductivity soil layer, a biotic barrier layer of large-sized cobbles may be needed above it. Where the type of waste may create gases, soil or geosynthetic vent structures would need to be included.

Settlement and subsidence should be evaluated for all covers and accounted for in the final cover plans. The current operating procedures for RCRA Subtitle C facilities (e.g., banning of liquids and partially filled drums of liquids) usually do not present major settlement or subsidence issues. For RCRA Subtitle D facilities, however, the normal decomposition of the waste will invariably result in settlement and subsidence. Settlement and subsidence can be significant, and special care may be required in designing the final cover system. The cover design process should consider the stability of all the waste layers and their intermediate soil covers, the soil and foundation materials beneath the landfill site, all the liner and leachate collection systems, and all the final cover components. When a significant amount of settlement and subsidence is expected within two to five years of closure, an interim cover that protects human health and the environment might be proposed. Then, when settlement/subsidence is essentially complete, the interim cover could be replaced or incorporated into a final cover.

Low Hydraulic Conductivity Layer

The function of the composite low hydraulic conductivity layer, composed of soil and a geomembrane, is to prevent moisture movement downward from the overlying drainage layer.

Compacted Soil Component

EPA recommends that a test pad be constructed before the low hydraulic conductivity soil layer is put in place to demonstrate that the compacted soil component can achieve a maximum hydraulic conductivity of 1×10^{-7} cm/s. To ensure that the design specifications are attainable, a test pad uses the same soil, equipment, and procedures to be used in constructing the low hydraulic conductivity layer. For Subtitle D facilities, the test fill should be constructed on part of the solid waste material to determine the impact of compacting soil on top of less resistive municipal solid waste.

The low hydraulic conductivity soil component placed over the waste should be at least 60-cm (24-in.) deep; free of detrimental rock, clods, and other soil debris; have an upper surface with a 5% maximum slope; and be below the maximum frost line. The surface should be smooth so that no small-scale stress points are created for the geomembrane.

In designing the low hydraulic conductivity layer, the causes of failure — subsidence, desiccation cracking, and freeze/thaw cycling — must be considered. Most of the settling will have taken place by the time the cover is put into place, but there is still a potential for further subsidence. Although estimating this potential is difficult, information about voids and compressible materials in the underlying waste will aid in calculating subsidence.

A soil with low cracking potential should be selected for the soil component of the low hydraulic conductivity layer. The potential for desiccation cracking of compacted clay depends on the physical properties of the compacted clay, its moisture content, the local climate, and the moisture content of the underlying waste.

Because freeze/thaw conditions can cause soil cracking, lessen soil density, and lessen soil strength, this entire low hydraulic conductivity/geomembrane layer should be below the depth of the maximum frost penetration. In northern areas, then, the maximum depth of the top vegetation/soil layer would be greater than the recommended minimum of 60 cm (24 in.).

Penetrating this low hydraulic conductivity/geomembrane soil layer with gas vents or drainage pipes should be kept to a minimum. Where a vent is necessary, there should be a secure, liquid-tight seal between the vent and the geomembrane. If settlement or subsidence is a major concern, this seal must be designed for flexibility to allow for vertical movement.

Geomembrane

The geomembrane placed on the smooth, even, low hydraulic conductivity layer should be at least 0.5-mm (20-mils) thick. The minimum slope surface should be 3% after any settlement of the soil layer or sub-base material. Stress situations such as bridging over subsidence and friction between the geomembrane and other cover components (i.e., compacted soil, geosynthetic drainage material, etc.), especially on side slopes, will require special laboratory tests to ensure that the design has incorporated site-specific materials.

Drainage Layer

The drainage layer should be designed to minimize the time the infiltrated water is in contact with the bottom, low hydraulic conductivity layer and, hence, to lessen the potential for the water to reach the waste (see Figure 14.1). Water that filters through the top layer is intercepted and rapidly moved to an exit drain, such as by gravity flow to a toe drain.

If the granular material in the drainage layer is sand, the minimum requirements are that it should be at least 30-cm (12-in.) deep with a hydraulic conductivity of 1×10^{-2} cm/s or greater. Drainage pipes should not be placed in any manner that would damage the geomembranes.

If geosynthetic materials are used in the drainage layer, the same physical and hydraulic requirements should be met, e.g., equivalency in hydraulic transmissivity, longevity, compatibility with geomembrane, compressibility, conformance to surrounding materials, and resistance to clogging. Geosynthetic materials are gaining increased use and understanding of their performance. Manufacturers are also continuing to improve the basic resin properties to improve their long-term durability. The net result is that organizations such as the American Society of Testing Materials (ASTM) and the Geosynthetic Research Institute (GRI),

Drexel University, Philadelphia, PA, are continually developing new evaluation procedures to better correlate with design and field experiences.

Between the bottom of the top-layer soil and the drainage-layer sand, a granular or geosynthetic filter layer should be included to prevent the drainage layer from clogging the top-layer fines. The criteria established for the grain size of granular filter sand are designed to minimize the migration of fines from the overlying top layer into the drainage layer. (For information on filter criteria, refer to the EPA Technical Guidance Document.[1]) ASTM test procedures have also been established to evaluate particulate clogging potential of geosynthetics.

Vegetation/Soil Top Layer

Vegetation Layer

The upper layer of the two-component top layer (Figure 14.1) should be vegetation (or another surface treatment) that will allow runoff from major storms while inhibiting erosion. Vegetation over soil (part of which is topsoil) is the preferred system, although, in some areas, vegetation may be unsuitable.

The temperature- and drought-resistant vegetation should be indigenous; have a root system that does not extend into the drainage layer; need no maintenance; survive in low-nutrient soil; and have sufficient density to control the rate of erosion to the recommended level of less than 5.5 MT/ha/yr (2 ton/acre/yr).

The surface slope should be the same as that of the underlying soils, at least 3% but no greater than 5% after consolidation. To support the vegetation, this top layer should be at least 60-cm (24-in.) deep and include at least 15-cm (6-in.) of topsoil. To help the plant roots develop, this layer should not be compacted. In some northern climates, this top layer may need to be more than the minimum 60-cm (24-in.) to ensure that the bottom low hydraulic conductivity layer remains below the frost zone.

Where vegetation cannot be maintained, particularly in arid areas, other materials should be selected to prevent erosion and to allow for surface drainage. Asphalt and concrete are apt to deteriorate because of thermal-caused cracking or deform because of subsidence. Therefore, a surface layer 13 to 25-cm (5 to 10-in.) deep of 5 to 10-cm (2 to 4-in.) stones or cobbles would be more effective. Although cobbles are a one-way valve and allow rain to infiltrate, this phenomenon would be of less concern in arid areas. In their favor, cobbles resist wind erosion well.

Soil Layer

The soil in this 60-cm (24-in.) top layer should be capable of sustaining nonwoody plants, have an adequate water-holding capacity, and be sufficiently deep to allow for expected, long-term erosion losses. A medium-textured soil such as a loam would fit these requirements. If the landfill site has sufficient topsoil, it should be stockpiled during excavation for later use.

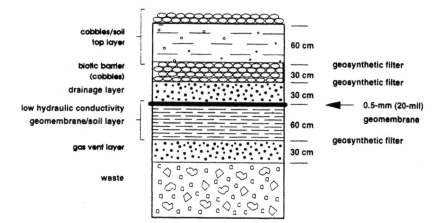

Figure 14.2 EPA recommended landfill cover with options.

The final slopes of the cover should be uniform and at least 3%, and should not allow erosion rills and gullies to form. Slopes greater than 5% will promote erosion unless controls are built in to limit erosion to less than 5.5 MT/ha/yr (2 ton/acre/yr). The U.S. Department of Agriculture's (USDA's) Universal Soil Loss Equation is recommended as the tool to evaluate erosion potential.

Optional Layers

Although other layers may be needed on a site-specific basis, the common optional layers are those for gas vents and for a biotic barrier layer (Figure 14.2).

Gas Vent Layer

The gas vent layer should be at least 30-cm (12-in.) thick and be above the waste and below the low hydraulic conductivity layer. Coarse-grained porous material, similar to that used in the drainage layer or equivalent-performing synthetic material, can be used.

The perforated, horizontal venting pipes should channel gases to a minimum number of vertical risers located at a high point (in the cross section) to promote gas ventilation. To prevent clogging, a granular or geotextile filter may be needed between the venting and the low hydraulic conductivity soil geomembrane layers.

As an alternative, vertical standpipe gas collectors can be built up as the landfill is filled with waste. These standpipes, which may be constructed of concrete, can be 30 cm (12 in.) or more in diameter and may also be used to provide access to measure leachate levels in the landfill.

Biotic Layer

Plant roots or burrowing animals (collectively called biointruders) may disrupt the drainage and the low hydraulic conductivity layers to interfere with the drainage capability of the layers. A 90-cm (3-ft) biotic barrier of cobbles directly beneath the top vegetation layer may stop the penetration of some deep-rooted plants and the invasion of burrowing animals. Most research on biotic barriers has been done in, as is applicable to, arid areas. Geosynthetic products that incorporate a time-released herbicide into the matrix or on the surface of the polymer may also be used to retard plant roots. The longevity of these products requires evaluation if the cover system is to serve for longer than 30 to 50 years.

SUBTITLE D COVERS

The cover system in nonhazardous waste landfills (Subtitle D) will be a function of the bottom liner system and the liquids management strategy for the specific site. If the bottom liner system contains a geomembrane, then the cover system should contain a geomembrane to prevent the "bathtub" effect. When the bottom liner is less permeable than the cover system, e.g., geomembrane on the bottom and natural soil on the top, the facility will "fill up" with infiltration water (through the cover) unless an active leachate removal system is in place. Likewise, if the bottom liner system is a natural soil liner, then the cover system barrier should be hydraulically equivalent to or less than the bottom liner system. A geomembrane used in the cover will prevent the infiltration of moisture to the waste below and may contribute to the collection of waste decomposition gases, therefore necessitating a gas-vent layer.

There are at least two options to consider under a liquids management strategy: mummification and recirculation. In the mummification approach, the cover system is designed, constructed, and maintained to prevent moisture infiltration to the waste below. The waste will eventually approach and remain in a state of "mummification" until the cover system is breached and moisture enters the landfill. A continual maintenance program is necessary to maintain the cover system in a state of good repair so that the waste does not decompose to generate leachate and gas.

The recirculation concept results in the rapid physical, chemical, and biological stabilization of the waste. To accomplish this, a moisture balance is maintained within the landfill that will accelerate these stabilization processes. This approach requires geomembranes in both the bottom and top control systems to prevent leachate from getting out and excess moisture from getting in. In addition, the system needs a leachate collection and removal system on the bottom and a leachate injection system on the top, maintenance of this system for a number of years (depending on the size of the facility), and a gas collection system to remove the waste decomposition gases. In a modern landfill facility, all of these elements, except the leachate injection system, would probably be available. The benefit of this approach is that, after stabilization, the facility should not require further maintenance. A more important advantage is that the decomposed and stabilized waste may be removed and used like compost, the plastics and metals could be recycled, and the site used again. If properly planned and operated in this manner, several cells could serve all of a community's waste management needs.

A natural soil material may be used in a cover system when the bottom liner system is also natural soils and the regulatory requirements will permit. A matrix of soil characteristics (using either USDA or USCS) and health, aesthetics, and site usage characteristics can be developed to provide information on which soil or combination of soils will be the most beneficial.

Health considerations demand the evaluation of each soil type to minimize vector breeding areas and attractiveness to animals. The soil should minimize moisture infiltration (best accomplished by fine grain soils) while allowing gas movement (coarse grain soils are best). This desired combination of seemingly opposite soil properties suggests a layered system. The soil should also minimize fire potential.

Aesthetic considerations include minimizing blowing of paper and other waste, controlling odors, and providing a sightly appearance. All landfill operators strive to be good neighbors and these considerations are very important for community relations.

The landfill site may be used for a variety of activities after closure. For this reason, cover soils should minimize settlement and subsidence, maximize compaction, assist vehicle support and movement, allow for equipment workability under all weather conditions, and allow healthy vegetation to grow. The future use of the site should be considered at the initial landfill design stages so that appropriate end-use design features can be incorporated into the cover during the active life of the facility.

CERCLA COVERS

The Superfund Amendments and Reauthorization Act of 1986 (SARA) adopts and expands a provision in the 1985 National Contingency Plan (NCP) that remedial actions must at least attain applicable or relevant and appropriate requirements (ARARs). Section 121(d) of CERCLA, as amended by SARA, requires attainment of federal ARARs and of state ARARs in state environmental or facility siting laws when the state requirements are promulgated, more stringent than federal laws, and identified by the state in a timely manner.

CERCLA facilities require information on whether or not the site is under the jurisdiction of RCRA regulations. The cover system design can then be developed based on appropriate regulations.

RCRA Subtitle C requirements for treatment, storage, and disposal facilities (TSDFs) will frequently be ARARs for CERCLA actions, because RCRA regulates the same or similar wastes as those found at many CERCLA sites, covers many of the same activities, and addresses releases and threatened releases similar to those found at CERCLA sites. When RCRA requirements are ARARs only the substantive requirements of RCRA must be met if a CERCLA action is to be conducted on site. Substantive requirements are those requirements that pertain directly to actions or conditions in the environment. Examples include performance standards for incinerators (40 CFR 264.343), treatment standards for land disposal of restricted waste (40 CFR 268), and concentration limits, such as maximum contaminant levels (MCLs). Onsite actions do not require RCRA permits or compliance with administrative requirements. Administrative requirements are those mechanisms that facilitate the implementation of the substantive requirements of a statute or regulation. Examples include the requirements for preparing a contingency plan, submitting a petition to delist a listed hazardous waste, recordkeeping, and consultations. CERCLA actions to be conducted off site must comply with both substantive and administrative RCRA requirements.

APPLICABILITY OF RCRA REQUIREMENTS

RCRA Subtitle C requirements for the treatment, storage, and disposal of hazardous waste are applicable for a Superfund remedial action if the following conditions are met:[2]

1. The waste is a RCRA hazardous waste, and either:
2. The waste was initially treated, stored, or disposed of after the effective date of the particular RCRA requirement
   or
   The activity at the CERCLA site constitutes treatment, storage, or disposal, as defined by RCRA.

For RCRA requirements to be applicable, a Superfund waste must be determined to be a listed or characteristic hazardous waste under RCRA. A waste that is hazardous because it once exhibited a characteristic (or a media containing a waste that once exhibited a characteristic) will not be subject to Subtitle C regulation if it no longer exhibits that characteristic. A listed waste may be delisted if it can be shown not to be hazardous based on the standards in 40 CFR 264.22. If such a waste will be shipped off site, it must be delisted through a rule-making process. To delist a RCRA hazardous waste that will remain on site at a Superfund site, however, only the substantive requirements for delisting must be met.

Any environmental media (i.e., soil or groundwater) contaminated with a listed waste is not a hazardous waste, but must be managed as such until it no longer contains the listed waste — generally when constituents from the listed waste are at health-based levels. Delisting is not required.

To determine whether a waste is a listed waste under RCRA, it is often necessary to know the source of that waste. For any Superfund site, if determination cannot be made that the contamination is from a RCRA hazardous waste, RCRA requirements will not be applicable. This determination can be based on testing or on best professional judgment (based on knowledge of the waste and its constituents).

A RCRA requirement will be applicable if the hazardous waste was treated, stored, or disposed of after the effective date of the particular requirement. The RCRA Subtitle C regulations that established the hazardous waste management system first became effective on November 19, 1980. Thus, RCRA regulations will not be applicable to wastes disposed of before that date, unless the CERCLA action itself constitutes treatment, storage, or disposal (see below). Additional standards have been issued since 1980; therefore, applicable requirements may vary somewhat, depending on the specific date on which the waste was disposed.

RCRA requirements for hazardous wastes will also be applicable if the response activity at the Superfund site constitutes treatment, storage, or disposal, as defined under RCRA. Because remedial actions frequently involve grading, excavating, dredging, or other measures that disturb contaminated material, activities at Superfund sites may constitute disposal, or placement, of hazardous waste. Disposal of hazardous waste, in particular, triggers a number of significant requirements, including closure requirements and land disposal restrictions, which require treatment of wastes prior to land disposal. (See Guides on Superfund Compliance with Land Disposal Restrictions, OSWER Directives 9347.3-01FS through 9237.3-06FS, for a detailed description of these requirements.)

EPA has determined that disposal occurs when wastes are placed in a land-based unit. However, movement within a unit does not constitute disposal or placement, and at CERCLA sites, an area of contamination (AOC) can be considered comparable to a unit. Therefore, movement within an AOC does not constitute placement.

Relevant and Appropriate RCRA Requirements

RCRA requirements that are not applicable may, nonetheless, be relevant and appropriate, based on site-specific circumstances. For example, if the source or prior use of a CERCLA waste is not identifiable, but the waste is similar in composition to a known, listed RCRA waste, the RCRA requirements may be potentially relevant and appropriate, depending on other circumstances at the site. The similarity of the waste at the CERCLA site to RCRA waste is not the only, nor necessarily the most important, consideration in the determination. An in-depth, constituent-by-constituent analysis is generally neither necessary nor useful, since most RCRA requirements are the same for a given activity or unit, regardless of the specific composition of the hazardous waste.

The determination of relevance and appropriateness of RCRA requirements is based instead on the circumstances of the release, including the hazardous properties of the waste, its composition and matrix, the characteristics of the site, the nature of the release or threatened release from the site, and the nature and purpose of the requirement itself. Some requirements may be relevant and appropriate for certain areas of the site, but not for other areas. In addition, some RCRA requirements may be relevant and appropriate at a site, while others are not, even for the same waste. For example, at one site minimum technology requirements may be considered relevant and appropriate for an area receiving waste because of the high potential for migration of contaminants in hazardous levels to groundwater, but not for another area that contains relatively immobile waste. Land disposal restrictions at the same site may not be relevant and appropriate for either area because the required treatment technology is not appropriate, given the matrix of the waste. Only those requirements that are determined to be both relevant and appropriate must be attained.

State Equivalency

A state may be authorized to administer the RCRA hazardous waste program in lieu of the federal program provided the state has equivalent authority. Authorization is granted separately for the basic RCRA Subtitle C program, which includes permitting and closure of TSDFs; for regulations promulgated pursuant to the Hazardous and Solid Waste Amendments (HSWA), such as land disposal restrictions; and for other programs, such as delisting of hazardous wastes. If a site is located in a state with an authorized RCRA program, the state's promulgated RCRA requirements will replace the equivalent federal requirements as potential ARARs.

An authorized state program may also be more stringent than the federal program. For example, a state may have more stringent test methods for characteristic wastes, or may list more wastes as hazardous than the federal program does. Therefore, it is important to determine whether laws in an authorized state go beyond the federal regulations.

Closure

For each type of unit regulated under RCRA, Subtitle C regulations contain standards that must be met when a unit is closed. For treatment and storage units, the closure standards require that all hazardous waste and hazardous waste residues be removed. In addition to the option of closure by removal, called clean closure, units such as landfills, surface impoundments, and waste piles may be closed as disposal or landfill units with waste in place, referred to as landfill closure. Frequently, the closure requirements for such land-based units will be either applicable or relevant and appropriate at Superfund sites.

Applicability of Closure Requirements

The basic prerequisites for applicability of closure (or AOC) requirements are

1. the waste must be hazardous waste; and
2. the unit must have received waste after the RCRA requirements became effective, either because of the original date of disposal or because the CERCLA action constitutes disposal.

When RCRA closure requirements are applicable, the regulations allow only two types of closure:

- Clean Closure. All waste residues and contaminated containment system components (e.g., liners), contaminated subsoils, and structures and equipment contaminated with waste leachate must be removed and managed as hazardous waste or decontaminated before the site management is completed [see 40 CFR 264.111,264.228(a)].

- Landfill Closure. The unit must be capped with a final cover designed and constructed to:
  - Provide long-term minimization of migration of liquids.
    - Function with minimum maintenance.
    - Promote drainage and minimize erosion.
    - Accommodate settling and subsidence.
    - Have a hydraulic conductivity less than or equal to any bottom liner system or natural subsoils present.

Clean closure standards assume the site will have unrestricted use and require no maintenance after the closure has been completed. These standards are often referred to as the "eatable solid, drinkable leachate" standards. In contrast, disposal or landfill closure standards require post-closure care and maintenance of the unit for at least 30 years after closure. Post-closure care includes maintenance of the final cover, operation of a leachate and removal system, and maintenance of a groundwater monitoring system [see 40 CFR 264.117, 264.228(b)].

EPA has prepared several guidance documents on closure and final covers.[1,3] These guidance documents are not ARARs, but are to be considered for CERCLA actions and may assist in complying with these regulations. The performance standards in the regulation may be attained in ways other than those described in guidance, depending on the specific circumstances of the site.

Relevant and Appropriate Closure Requirements

If they are not applicable, RCRA closure requirements may be determined to be relevant and appropriate. There is more flexibility in designing closure for relevant and appropriate requirements because the Agency has the flexibility to determine which requirements in the closure standards are relevant and appropriate. Under this scenario, a hybrid closure is possible. Depending on the site circumstances and the remedy selected, clean closure, landfill closure, or a combination of requirements from each type of closure may be used.

The proposed revisions to the National Contingency Plan (NCP) discuss the concept of hybrid closure (53 CFR 51446). The NCP illustrated the following possible hybrid closure approaches:

- Hybrid-Clean Closure. Used when leachate will not impact the groundwater (even though residual contamination and leachate are above health-based levels) and contamination does not pose a direct contact threat. With hybrid-clean closure:
  - No covers or long-term management are required.
  - Fate and transport modeling and model verification are used to ensure that groundwater is usable.
  - A property deed notice is used to indicate the presence of hazardous substances.

- Hybrid-Landfill Closure. Used when residual contamination poses a direct contact threat, but does not pose a groundwater threat. With hybrid-landfill closure:
  - Covers, which may be permeable, are used to address the direct contact threat.
  - Limited long-term management includes site and cover maintenance and minimal groundwater monitoring.
  - Institutional controls (e.g., land-use restrictions or deed notices) are used as necessary.

The two hybrid closure alternatives are constructs of applicable laws but are not themselves promulgated at this time. These alternatives are possible when RCRA requirements are relevant and appropriate, but not when closure requirements are applicable.

ALTERNATIVE COVER DESIGNS

The hazardous waste landfill cover designs developed and published by the U.S. Environmental Protection Agency (EPA) are generic in nature and intended to meet the regulatory criteria for covers on a national basis. This section reviews these designs to determine alternatives that would be acceptable to the regulatory community. This discussion will also review designs being considered by other agencies, such as the Nuclear Regulatory Commission (NRC).

SUBTITLE C

The basic acceptable generic design for hazardous waste landfills incorporates natural soil (clay), geomembrane, drainage, and vegetation layers. This generic design, however, does not take into account site-specific concerns, such as siting in arid areas where rainfall is very low. Under these conditions, a barrier layer composed of both a natural soil (clay) and a geomembrane layer probably would not be effective. The natural soil layer is designed to be placed "wet of optimum" to achieve the minimum hydraulic conductivity. When placed in a relatively dry environment, this layer will dry and crack, making it less effective. In selected cases, the newer bentonite blankets (GCLs) may be an acceptable alternative.

From a technical standpoint, the geomembrane may be the only barrier necessary. Site-specific considerations such as settlement/subsidence, environmental exposure, and other physical conditions may influence the thickness of the geomembrane required.

In selected cases, a vegetation layer alone may be demonstrated [via the Hydrologic Evaluation of Landfill Performance (HELP) model[4]] to meet the criteria. In this case, a thicker soil layer may be required to assist in establishing the natural vegetation and to act as a storage reservoir for the infrequent but high intensity rainfall.

In summary, the design criteria were established for a national generic design. EPA is always interested in reviewing alternative designs that are innovative and utilize site-specific information. These alternative designs should be demonstrated to be equivalent in performance to the generic design proposed by EPA.

SUBTITLE D

While EPA has proposed some generic design considerations, Subtitle D facility designs will most likely be approved by individual states. Cover designs should be incorporated into the overall facility design, taking the bottom liner and liquids management strategy into account. Depending on site-specific considerations, designs based on natural soils as well as designs that resemble multi-layer Subtitle C designs will be developed.

Municipal solid waste landfills usually require a daily cover of natural soils or other alternative materials. One possible use for post-consumer paper or unsaleable glass or glass culls may be for daily cover. A product has been made from shredded paper mixed with other proprietary ingredients that can be blown onto the surface of the waste to meet the requirements of daily cover. Foams and other materials have also been developed and evaluated for performance. Each of these materials has to be evaluated economically for site-specific use but may have advantages technically. For example, if the liquids management strategy for a landfill includes leachate recirculation, blown-on materials will lose their barrier qualities as soon as the next layer is placed in the facility. Natural soils, on the other hand, do tend to act as barriers, which may cause leachate to seep out the side of the final cover.

Whatever alternate materials are used, they should be demonstrated to meet the technical requirements for daily cover.

CERCLA

CERCLA or Superfund cover designs are more complex from the standpoint of jurisdiction, where ARARs play an important part in selecting the final design. A multi-layer cover system may be most environmentally desirable; however, other site-specific considerations may allow other types of designs. For example, early CERCLA covers have been constructed by regrading existing cover material, and adding small amounts of cover soil (usually about 6 inches) and, in some cases, a rock armor. Due to the inequality between RCRA and CERCLA and with the ARARs ruling, compliance with a RCRA multi-layer has been more acceptable. Site-specific design changes have been approved after they were demonstrated to meet the intent of the regulations.

OTHER COVER DESIGNS

The Department of Energy (DOE) and the Nuclear Regulatory Commission (NRC) are both considering cover designs for landfills containing low level

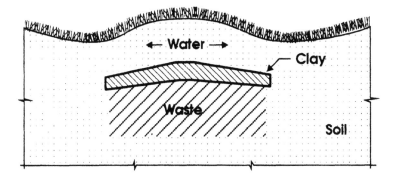

Figure 14.3 Resistive layer barrier.

radioactive wastes. In general, these designs are comparable to the EPA's multi-layer design, with some notable exceptions. One of the main criteria differences is based on the fact the DOE and NRC designs have a 10,000 year minimum service life due to the type of waste they are covering. The long-term nature of their designs has minimized the use of geosynthetics, since geosynthetics are thought to have a finite service life.

DOE will soon publish results of a study designed to develop an all natural soil cover system with a long service life.[4] The study considered what type of soil would best qualify for each design aspect. A matrix was developed from which completed matrix designs will be proposed.

NRC also has been reviewing conceptual designs that use natural soils and have long life. Three cover designs are currently under investigation:

a. resistive layer barrier,
b. conductive layer barrier, and
c. bioengineering barrier.[2]

These designs are being assessed in large (21 × 14 × 3 m [70 × 45 × 10 in.] each) lysimeters in Beltsville, MD by the Nuclear Regulatory Commission. The resistive layer barrier, shown in Figure 14.3, consists of compacted natural soils or clay. The resistive layer depends on the low hydraulic conductivity of the compacted layer to minimize any potential moisture interaction with the waste.

The conductive layer barrier, shown in Figure 14.4, makes use of the capillary barrier phenomena to increase the moisture content above the interface and to divert water away from and around the waste.[6] The capillary barrier is established when coarse grain soils are sandwiched between fine grain sediments. Experiments have shown that the greater the contrast in the permeability between the two layers, the more effective the barrier. A second fine grain soil layer would direct water away from the gravel layer under saturated conditions.

It should be noted that NRC considers these two conceptual designs unacceptable where appreciable subsidence may take place.[5] This failure potential in the above two designs necessitated the development of an easily reparable surface

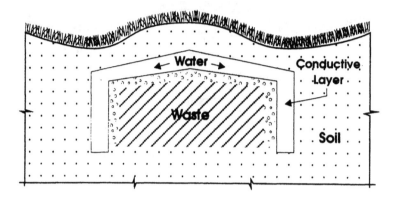

Figure 14.4   Conductive layer barrier.

barrier to be used until major settlement/subsidence time period, after which a more permanent barrier could be installed. The bioengineering management cover system[5] was the result. This cover system, shown in Figure 14.5, utilizes a combination of engineered enhanced runoff and stress vegetation, e.g., Pfitzer junipers, growing in an overdraft condition to control deep water percolation

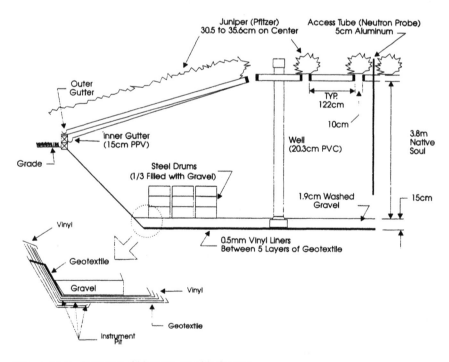

Figure 14.5   Sideview of bioengineered lysimeter.

through cover systems. Stress vegetation are grasses, trees, and shrubs that can survive when under stress, such as lack of water. Early results from the field indicate this system to be very effective in controlling liquid movement into or out of the waste management unit.

GAS GENERATION

The information in this chapter applies mainly to Subtitle D landfills. Hazardous waste landfills (Subtitle C) do not usually contain significant amounts of organic materials and, thus, normally have a minimal gas management system as a component of the final cover.

Gas generation in a landfill system poses several problems. If allowed to accumulate, gas is an explosion hazard. It also provides stress to vegetation by lowering the oxygen content available at the roots, severely affecting the ability of the cover to support vegetation. In the absence of adequate corridors for the gas to escape, gas pressures can increase sufficiently to physically disrupt the cover system as well, generating large cracks and rupturing the geomembrane. Other problems include odor, toxic vapors, and uncontrolled gas migration which can cause deterioration of nearby property values.

Gas generation is a product of anaerobic decomposition of organic materials placed in the landfill. The decomposition can be described by the reaction:

$$C_aH_bO_cN_dS_e \rightarrow CH_4 + wCO_2 + xN_2 + yNH_3 + zH_2S + humus^1$$

The composition of landfill gas generally is about 50% methane, 40% carbon dioxide, and 10% other gases including nitrogen products. This particular mix of gases generally will not occur until after the landfill becomes anaerobic. During the first year after the materials are placed in the landfill, the gas is predominantly carbon dioxide and is unsuitable for recovery and use. After the methane content rises, the gas can be mined as a fuel or energy source. However, the BTU value of landfill gas is about half that of natural gas and, therefore, is generally too low to substitute directly for natural gas. Landfill gas requires purification and is frequently used in conjunction with natural gas.

Waste decomposition rates and hence gas production rates are moisture dependent. Highest gas production rates occur at moisture contents ranging from 60 to 80% of saturation. In modern landfill design, infiltration of water into the waste is restricted to a practical minimum; therefore, optimum moisture contents may never be achieved. Consequently, gas production rates may be much lower than anticipated, decreasing the attractiveness of gas recovery systems. To maximize gas production, strategies such as leachate recirculation should be employed to distribute bacteria, nutrients, and moisture more uniformly. Typical gas production rates from wet, anaerobic wastes are about 20 to 50 mL/kg/day. These high production rates will continue for centuries because of quantity of material and resistance of some material for biodegration.

GAS MIGRATION

Gas migrates from landfills through two mechanisms — convection and diffusion. Convection is transported induced by pressure gradients formed by gas production in layers surrounded by low hydraulic conductivity or saturated layers. Convection also results from buoyancy forces because methane is lighter than carbon dioxide and air.

Diffusion is the transport of materials induced by concentration gradients. Anaerobic decomposition produces a gas mixture with concentrations of methane and carbon dioxide that are much greater than those found in the surrounding air. Therefore, molecules of methane and carbon dioxide will diffuse from the landfill gas to the air in accordance with Fick's law. Diffusion plays a much smaller role in gas migration than convection.

Many factors affect gas migration. Some of the more important factors are the landfill design, including refuse cell construction; final cover design; and incorporation of gas migration control measures. Low hydraulic conductivity soil layers and geomembranes are very effective barriers to gas migration. Sand and gravel layers and void spaces provide effective corridors for channeling gas migration. Other channels affecting migration are cracks and fissures between and in lifts of waste or soil due to differential settlement and subsidence.

Other factors affecting gas migration include the gas production rate, the presence of natural and artificial conduits and barriers adjacent to the landfill, and climatic and seasonal variations in site conditions. High gas production rates increase migration. Corridors at the site adjacent to a landfill such as water conduits, drain culverts, buried lines, and sand and gravel lenses, promote uncontrolled migration from the site. Barriers can include clay deposits; high or perched water tables, roads; and compacted, low hydraulic conductivity soils. Environmental variations can result from the intermittent occurrence of saturated or frozen surface soils, which seals the surface and promotes lateral migration. Barometric pressure changes also affect the rate of gas release to the surface. Seasonal changes in moisture content can change the gas production rate and, therefore, the extent and quantity of migration.

GAS CONTROL STRATEGIES

Two gas control strategies — passive and active — are available, and may be used at any facility. Passive systems provide corridors to intercept lateral gas migration and channel the gas to a collection point or a vent. These systems use barriers to prevent migration past the interceptors and the perimeter of the landfill. Active systems generate a zone of negative pressure to increase the pressure gradient and, consequently, the flow toward the zone. Active systems also can be used to create a zone of high pressure to prevent gas migration toward the zone.

Typical passive systems are shown in Figures 14.6, 14.7, 14.8, and 14.9. Figure 14.6 shows a gas-vent layer used in conjunction with a composite liner

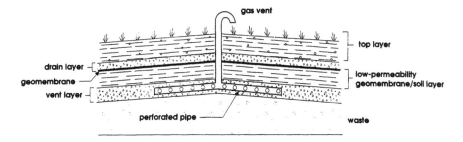

Figure 14.6 Cover with gas vent outlet and vent layer.

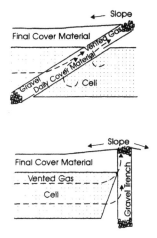

Figure 14.7 Gravel vent and gravel-filled trench used to control lateral gas movement in a sanitary landfill.

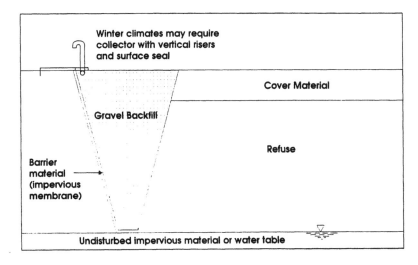

Figure 14.8 Typical trench barrier system.

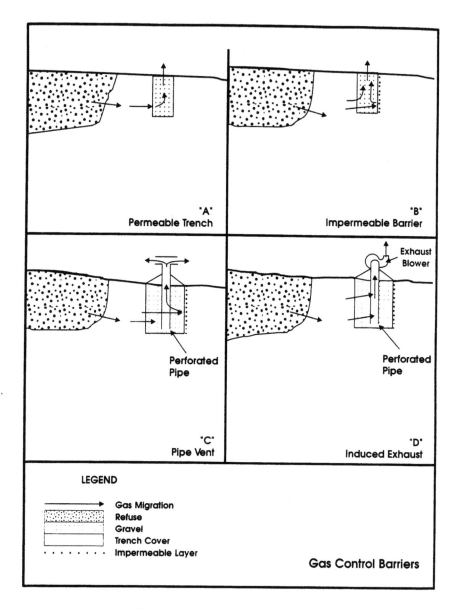

Figure 14.9 Gas control barriers.

and vent in the cover system.[1] The composite liner prevents uncontrolled vertical migration, while the gas-vent layer intercepts all vertical migration and directs it to the vent. Figure 14.7 shows a gravel vent that runs diagonally down through the waste material. The gravel vent intercepts both vertical and lateral migration and channels it to the surface. Figures 14.7, 14.8, and 14.9 show gravel-filled

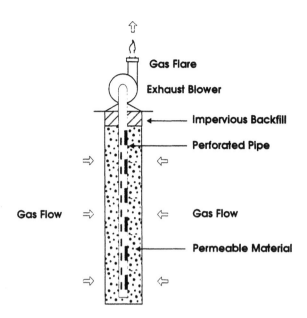

Figure 14.10 Gas extraction well for landfill gas control.[8-12]

trenches.[8-10] The trenches intercept lateral migration and direct the gas to the surface where it is vented or extracted. Gravel-filled trenches on the perimeter of the landfill are often used with an impermeable barrier on the outer side of the trench to prevent migration from the trench to the surrounding area. These systems often extend from the surface down to a low hydraulic conductivity soil layer or other barrier such as the water table or a geomembrane. The systems may be as deep as the bottom of the landfill, or even lower if outside the landfill. Extreme care should be taken in the design of all of these vent systems to prevent them from being a source of infiltration through the cover. Improper design could allow the vent to intercept surface runoff and pipe additional infiltration into the leachate collection system.

Typical active systems are shown in Figures 14.9, 14.10, and 14.11. All three figures show gas extraction wells using exhaust blowers. The well is placed in a gravel vent or gravel-filled trench located in the waste cell (Figures 14.10 and 14.11) or along its perimeter (Figure 14.9). The gravel vent is sealed to prevent the well from drawing air from the surface and destroying the suction (zone of negative pressure) needed to draw gas to the well. The seal also prevents infiltration of surface water. Impermeable barriers in the cover and perimeter walls increase the efficiency of gas extraction wells since they restrict inflow of air that would dissipate the suction. In addition, it reduces the number of wells needed and increases the heating value of the gas collected. Typically, gas extraction wells do not extend to the bottom of the landfill since the suction is able to draw gas from a sizable zone beyond the gravel fill.

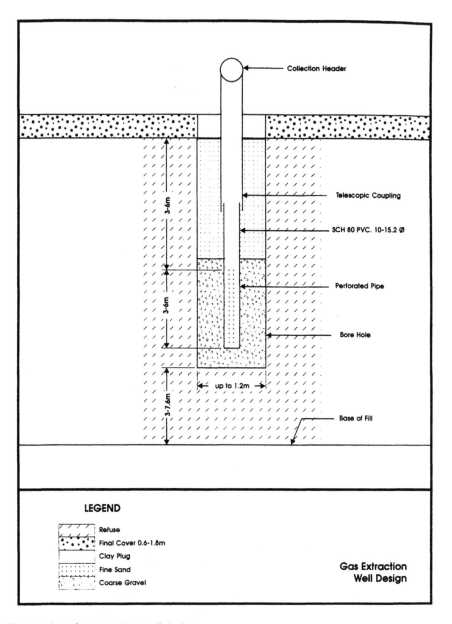

Figure 14.11 Gas extraction well design.

REFERENCES

1. U.S. EPA, 1989. Final covers on hazardous waste landfills and surface impoundments. Office of Solid Waste and Emergency Response Technical Guidance Document EPA 530/SW-89/047, Risk Reduction Engineering Laboratory, Cincinnati, OH.

2. U.S. EPA, 1989. RCRA ARARs: focus on closure requirements. Office of Solid Waste and Emergency Response Directive 9234.2-04FS, Office of Solid Waste and Emergency Response, Washington, D.C.

3. U.S. EPA, 1978. Closure and post-closure standards. Draft RCRA Guidance Manual for Subpart G. EPA 530/SW-78/010. Office of Solid Waste and Emergency Response, Washington, D.C.

4. U.S. Army Corps of Engineers, Identification and ranking of soils for UMTRA and LLW disposal facility covers. Waterways Experiment Station. Unpublished.

5. O'Donnell, E., Ridky, R.W., and Schultz, R.K., 1990. Control of water infiltration into near surface LLW disposal units. Progress report on field experiments at a humid region site, Beltsville, MD. Waste Management `90 Tucson, Arizona.

6. Zunker, J.F., 1930. Das verhalten des bodens zum wasser In: E. Blanck, Ed. Handbuch Der Bodenlehr. V. 6. Berlin: Verlag von Julius Springer, pp. 66–220.

7. U.S. EPA, 1994. Hydrologic evaluation of landfill performance version 3.05. User's Guide for Version 3. EPA/600/R-94/168a and b USEPA, Risk Reduction Engineering Laboratory, Cincinnati, OH, September, 1994.

8. Lutton, R.J., Regan, G.L., and Jones, L.W., 1979. Design and construction of covers for solid waste landfills. EPA-600/2-79/165, U.S. EPA Municipal Environmental Research Laboratory, Cincinnati, OH.

9. Shafer, R.A., Renta-Babb, A., Bandy, J.T., Smith, E.D., and Malone, P., 1984. Landfill gas control at military installations. Technical Report N-173. U.S. Army Engineer Construction Engineering Research Laboratory, Champaign, IL.

10. McAneny, C.C., Tucker, P.G., Morgan, J.M., Lee, C.R., Kelley, M.F., and Horz, R.C., 1985. Covers for uncontrolled hazardous waste sites. EPA/540/2-85/002, U.S. EPA Hazardous Waste Engineering Research Laboratory, Cincinnati, OH.

11. Brunner, D.R. and Keller, D.J., 1971. Sanitary landfill design and operation. SW-66TS. U.S. Environmental Protection Agency, Cincinnati, OH.

12. Rovers, F.A., Tremblay, J.J., and Mooij, H., 1977. Procedures for landfill gas monitoring and control. EPS4-EC-77-4, Waste Management Branch, Environment Canada, Ottawa.

CHAPTER 15

Design of Subtitle D and Subtitle C Landfill Containment Systems

Robert M. Koerner

INTRODUCTION

Currently, the federal U.S. Environmental Protection Agency (EPA) regulations for nonhazardous (Subtitle D) and hazardous (Subtitle C) landfills are in place and the various states are active in bringing their regulations into compliance. This chapter reviews the current status for liner systems beneath the waste. This involves both the barrier system(s) and the leachate collection system(s). In presenting this information, it is important to recognize that regulations generally allow for replacement materials, as long as they can be justified on the basis of technical equivalency. This allows for the substitution (or augmentation) of geosynthetics for natural soil materials in many instances. The various situations are described in this chapter.

An important aspect of landfills beyond the liner and cover systems is the liquids management scheme that is utilized at the site. Various options of leachate collection and removal systems and the implications on design are described as is leachate recirculation. The future status of the design of landfill liner systems is suggested, whereby fine-tuning of the materials involved will be the focus, along with data collection regarding field performance. This phase in the genesis of landfill containment systems has recently been initiated via a major U.S. EPA cooperative agreement. Work is actively ongoing in this regard.

Since the U.S. Environmental Protection Agency regulations on the disposal of solid waste is enormous, only those aspects of relevance to the design of landfill liners will be discussed here. Landfill covers are treated separately; see Landreth.[1]

Applicable Legislation

Solid waste materials that are subsequently landfilled are subdivided into nonhazardous and hazardous categories by virtue of the chemical characteristics of the site-specific leachate. Leachate is the liquid contained in the waste and the subsequent rainfall and snowmelt that flows through it. Obviously, the leachate takes on the characteristics of the solid waste and is therefore always waste-specific, i.e., there is no typical leachate. The site-specific leachate is analyzed by conventional analytical techniques (usually TCLP sampling, then mass spectroscopy, and gas chromatography analysis) and if all of the 800+ listed priority pollutants are less than stipulated EPA limits, the waste mass is considered nonhazardous. Its disposal then falls under EPA regulations 40 CFR Parts 257 and 258, Subtitle D, for solid waste disposal. Some salient points regarding Subtitle D liner systems are the following, which gives rise to the cross section shown in Figure 15.1(a).

- A leachate collection system should be located above the liner system.
- The leachate collection system should be capable of maintaining a leachate head of less than 12 in. (30 cm).
- The liner system should be a *single composite liner* (i.e., it is not required to have a double liner system with leak detection capability as it is with hazardous waste materials).
- The single composite liner must be a geomembrane placed over a compacted clay liner.
- The geomembrane must be 30 mils (0.75 mm) thick, unless it is HDPE. In the latter case it must be 60 mils (1.50 mm).
- The geomembrane must have "direct and uniform contact with the underlying compacted soil component." Furthermore, the term *intimate contact* is referenced in many regulations.
- The compacted clay liner beneath the geomembrane must be 24 in. (60 cm) thick and of a permeability of 2×10^{-7} ft/min (1×10^{-7} cm/s) or less. (Note that permeability will be used in this chapter rather than the more accurate term of hydraulic conductivity.)

If any one of the 800+ listed priority pollutants exceeds the stipulated EPA limits, the solid waste is considered hazardous. Its disposal then falls under EPA regulations 40 CFR 264.221, "Subtitle C," for solid waste disposal. Some salient points regarding Subtitle C liner systems are the following, which gives rise to the cross section shown in Figure 15.1(b).

- The leachate collection system should be capable of maintaining a leachate head of less than 12 in. (30 cm).
- A *double liner system* with leak detection capability between them is to be located directly beneath the leachate collection system.
- A leak detection system should be located between the two liners.
- Both leachate collection and leak detection systems should have at least 12-in. (30-cm) granular layers that are chemically resistant to the waste and leachate,

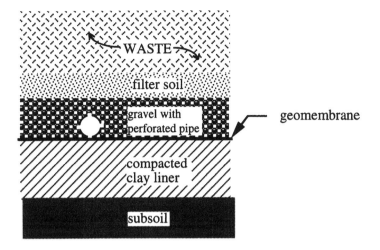

(a) Single composite liner, per
     Subtitle "D" regulations

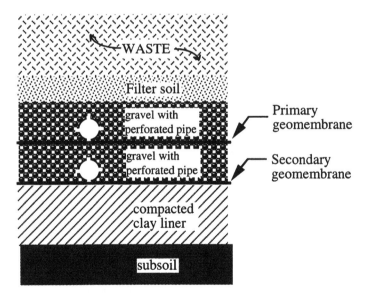

(b) Double liner system per
     Subtitle "C" regulations

Figure 15.1   Minimum technology guidance (MTG) liner systems for (a) nonhazardous and
(b) hazardous waste containment.

with a permeability not less than 2×10^{-2} ft/min (1×10^{-2} cm/s), or an equivalent synthetic drainage material, e.g., a geonet or geocomposite.
- The minimum bottom slope of the facility should be 0.5%.
- The leachate collection system should have a granular filter or synthetic geotextile above the drainage layer to prevent clogging.
- Both collection systems, when made of natural soils, should have a network of interconnected perforated pipes to remove the leachate; the pipes should have sufficient strength and chemical resistance to perform under anticipated landfill loads.
- By virtue of the leak detection rules contained in 40 CFR 260, 264, 265, 270, 271, a site-specific action leakage rate (ALR) must be set for each facility.
- A construction quality assurance (CQA) program must be developed to ensure that the constructed facility meets or exceeds all design criteria, plans, and specifications.
- The geomembranes must be 30 mils (0.75 mm) thick, unless it is HDPE. In the latter case it must be 60 mils (1.5 mm).
- The compacted clay liner beneath the secondary geomembrane must be 24 in. (60 cm) thick and of a permeability of 2×10^{-7} ft/min (1×10^{-7} cm/s), or less.

Leak Detection

The hallmark of the double liner system as shown in Figure 15.1(b) is that leakage passing through the primary liner can be monitored and quantified on a routine basis. The quantity is then compared to the action leakage rate (ALR) that is prescribed in the permit for the facility. If, indeed, leakage of leachate through the primary liner is noted above the ALR, corrective measures must be instituted. The prescribed corrective measures will be delineated in the response action plan (RAP), which necessarily accompanies the regulatory permit and also sets the site-specific ALR value. Such actions might be as follows:

- Continuous monitoring and tracking of the leakage rate.
- Chemical analysis of both the liquid in the leak detection system and the leachate in the primary collection system for comparative purposes.
- Placement of downstream monitoring wells (or additional monitoring wells).
- Secession of waste placement in the facility.
- Removal of waste from the facility to find and repair the leak(s).

The leak detection capabilities of a double liner system are so significant that at least eight states have adopted this type of strategy for nonhazardous waste, i.e., for municipal solid waste.
Since the above listed RAPs are quite costly on the part of the facility owner, some double liner systems are designed with a clay layer beneath the primary geomembrane. This clay layer can take the form of a compacted clay liner (CCL), or a geosynthetic clay liner (GCL). Details of these various materials will be given later.

Technical Equivalency

In federal regulations and in most (if not all) state regulations, material substitutions can be made if technical equivalency can be demonstrated. Thus, there are a number of possible substitutions that generally consist of replacing natural soil materials with a geosynthetic (polymeric) material such as:

- geotextiles for sand filters,
- geonets for gravel drains,
- GCLs for CCLs,
- geogrids for reinforcement of side slopes,
- various geosynthetics for daily cover, and
- geosynthetic erosion control materials for vegetative cover.

Such geosynthetic materials are easy to place, generally less expensive than natural soils, and invariably save air space resulting in greater availability of landfill volume for additional waste placement.

Geosynthetic Materials

There is a large and ever growing body of technical information on geosynthetics; for example, see Koerner.[2] For geosynthetics used in the liner of a solid waste landfill, the most important types are described as follows:

- Geomembranes (GM) are very low permeability, polymeric membrane liners or barriers used to contain the generated leachate and to control fluid migration.
- Geotextiles (GT) are planar, permeable, polymeric materials comprised solely of textiles used for filtration, drainage, separation, or reinforcement.
- Geosynthetic clay liners (GCL) are factory-manufactured hydraulic barriers consisting of a layer of bentonite clay or other very low permeability material, supported by geotextiles and/or geomembranes, being mechanically held together by adhesives, needling, or stitching.
- Geonets (GN) are net-like sets of interconnected polymer ribs used for the transmission of liquids in the plane of the structure.
- Geogrids (GG) are grid-like sets of interconnected polymer ribs used for reinforcement of soils or solid waste.

These geosynthetics are shown in Figure 15.2 for a double liner system in a possible (but clearly not typical) situation. Moving from the waste downward to the soil/rock subgrade, the components are as follows:

- solid waste in lifts with soil daily cover, or alternate daily cover materials (ADCM)
- leachate collection and removal systems consisting of a geotextile filter over sand or gravel

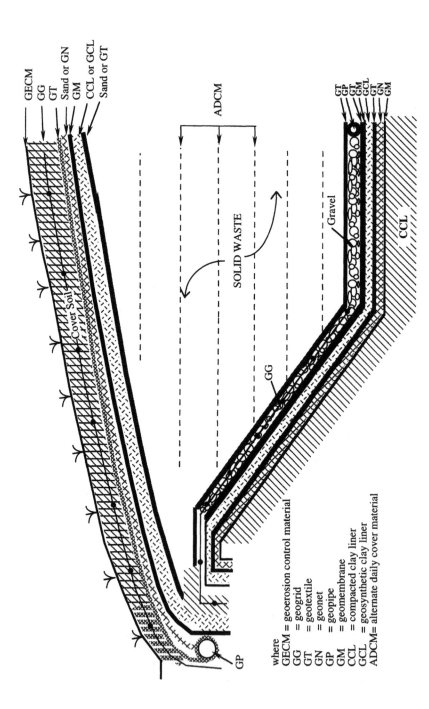

Figure 15.2 Various components used in double liner waste containment system.

where
GECM = geoerosion control material
GG   = geogrid
GT   = geotextile
GN   = geonet
GP   = geopipe
GM   = geomembrane
CCL   = compacted clay liner
GCL   = geosynthetic clay liner
ADCM= alternate daily cover material

- within the leachate collection layer is a perforated pipe system for leachate removal
- also within the leachate collection layer on side slopes a geogrid reinforcement layer is sometimes necessary
- the primary liner system consisting of a geomembrane placed directly on a geosynthetic clay liner, i.e., a GM/GCL composite
- the leak detection system consisting of a geotextile bonded to a geonet, i.e., a GT/GN composite
- the secondary liner system consisting of a geomembrane on a compacted clay liner, i.e., a GM/CCL composite and
- the underlying subgrade soil or rock material.

SINGLE LINER SYSTEMS

As shown in Figure 15.1(a), a single composite liner system is the minimum technology guidance (MTG) for nonhazardous waste in federal regulations and in most state regulations.

Barrier Materials

The containment of leachate (or any other liquid) in a landfill comes about by a barrier material that is chemically resistant to the liquid to be contained. Other than a naturally occurring deep geologic deposit, there are three engineered options:

- Compacted clay liners (CCLs) — These barriers are composed of clayey soil materials that are placed and compacted in layers called *lifts*. The materials used to construct CCLs include natural mineral materials (usually fine grained silt and clay soils), bentonite-soil blends and other materials. Many references on CCLs are available, see Daniel[3] and Goldman, et al.[4]
- Geomembranes (GMs) — Geomembranes are very low permeability polymer sheets in thicknesses ranging from 20 to 120 mils (0.50 to 3.0 mm). The most commonly used types for landfill barrier construction are HDPE, VFPE, PVC, FPP, and CSPE-R, see Koerner.[2]
- Geosynthetic Clay Liners (GCLs) — Geosynthetic clay liners are typically 10 mm thick layers of bentonite carried by geotextiles or a geomembrane and are very low in their permeability, typically $(1 \text{ to } 5) \times 10^{-9}$ cm/s, see Daniel and Boardman.[5]

Critical in the selection of a barrier material to the containment of leachate is its chemical resistance. This, of course, is interrelated to its long-term durability. Beyond this criterion, the required properties of the barrier material should be handled in the site specific engineering design.

Composite GM/CCL Liners

The concept of a composite liner is to place a geomembrane directly above a CCL. In this way leakage through a potential hole in the geomembrane imme-

Table 15.1 Generalized Leakage Rates Through Liners[14]

Type of liner	Leakage mechanism	Liquid height above the geomembrane, h_w			
		0.1 ft (0.03 m)	1 ft (0.3 m)	10 ft (3 m)	100 ft (30 m)
Geomembrane alone	Diffusion	0.001	0.1	10	30
(between two pervious	Small hole[a]	30	100	300	1,000
soil layers)	Large hole[a]	1,000	3,000	10,000	30,000
Composite liner	Diffusion	0.001	0.1	10	30
(good field conditions)	Small hole[a]	0.015	0.1	0.9	7.5
	Large hole[a]	0.02	0.15	1.1	8.5
Composite liner	Diffusion	0.001	0.1	10	30
(poor field conditions)	Small hole[a]	0.08	0.6	5.0	40
	Large hole[a]	0.1	0.7	6.0	50

Note: Values of leakage rate in gal/acre-day (figures to be multiplied by approximately 10 in order to obtain values in units of l/ha-day).

[a] Assumes 3 holes/ha (i.e., 1 hole/acre).

After Giroud and Bonaparte.[14]

diately confronts the clay, but only in the vicinity of the hole. Shown in Table 15.1 are the extremely low leakage rates through a composite liner, versus the geomembrane by itself. For example, at a liquid height of 1.0 ft (0.3 m) above a geomembrane with a small hole, the leakage rate is 100 gal/acre-day when it is on sand versus 0.1 gal/acre-day when it is on a clay liner. Thus the leakage is reduced by three orders of magnitude. This is the essence of a composite liner. Clearly, the concept of a composite liner allows the designer considerable security against leakage into the underlying material.

Composite GM/GCL Liners

Although significantly less study has been performed on GM/GCL composite liners they, too, can act in a manner similar to that described above. One concern, however, is the lateral transmission of leachate in the geotextile covering the GCL. This has recently been evaluated by Harpur, et al.[6] Using a radial transmissivity device, the flow rates in terms of transmissivity at the interface between a geomembrane and various GCLs were measured. Five different GCLs were used beneath a 60 mil (1.5 mm) HDPE geomembrane at two normal pressures. Figure 15.3 shows these results. GCL "A" has bentonite directly beneath the geomembrane and thus defines the lowest possible transmissivity. The other GCLs are slightly higher and respond somewhat to increasing pressure since each has a covering geotextile. Of significance, however, is that the predicted GM/CCL values are all higher than any of the measured GM/GCL combinations. Thus a GM/GCL composite liner will probably give equivalent, or better, performance than a GM/CCL composite liner in this regard.

Natural Soil Drainage Layers

Federal and state regulations generally require 12 in. (300 mm) of granular soil having a permeability of 1×10^{-2} cm/s, or higher, for leachate collection.

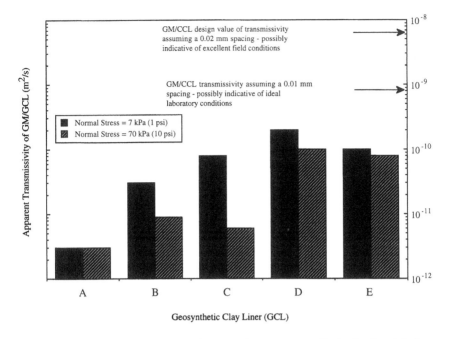

Figure 15.3 Apparent transmissivity of geomembrane-to-geosynthetic clay liner interface.

Two considerations must also be addressed when using natural soil drainage layers (coarse sands or gravels); one is that a perforated pipe network must be located in the drainage layer, and the second is that a filter must be placed over the drainage layer. The filter layer should be 6 in. (150 mm) of sand, appropriately designed considering the waste above it and the drainage layer below.

Geosynthetic Drainage Layers

Since replacement materials are allowed, if they are technically equivalent to the regulatory suggested materials, and for the practical reasons described earlier, geonet drains and geotextile filters are frequently substituted for natural soil leachate drainage systems. When using geonets, pipe systems are not necessary since flow rates in geonets are very rapid in comparison to soil drains. Also, the required geotextile filters are usually thermally bonded to the geonet in the factory, thereby avoiding a potentially weak shear surface. Such GN/GT composites are regularly used on side slopes for the additional reason of slope stability.

Geotextile Protection Layers

When gravel is used as the leachate collection drain, the adjacent geomembrane must be protected against puncture. The typical protection layer is a needle-punched, nonwoven geotextile of 12 to 18 oz/yd^2 (400 to 600 g/m^2) mass per unit area. The design of various protection materials is being actively pursued, see Narejo.[7]

Geogrid Reinforcement Layers

Utilizing geogrid or high strength geotextile reinforcement layers, the practices of veneer stability and landfill expansions often come about. *Veneer stability* consists of embedding a layer of geosynthetic reinforcement in the potentially unstable soil to assist (or completely sustain) the gravitational, and seismic seepage stresses that are mobilized. *Landfill expansions* consist of vertical and horizontal expansions of existing landfills and often use a layer of geogrid or high strength geotextile to support the new liner system against differential settlements in the underlying solid waste. For both of these applications the design methodology is reasonably well formulated, see Koerner.[2]

DOUBLE LINER SYSTEMS

The use of two separate liner systems with a leak detection layer between them as in Figures 15.1(b) and 15.2 is a very powerful liner strategy. Periodic monitoring of the quantity of liquid in the leak detection layer can give significant insight as to the performance of the primary liner above it and pose possible implications to the secondary liner below.

Primary Liner of a Double Liner System

The primary liner of a double liner system can be a single barrier material (GM, CCL, or GCL), however, federal regulations require at least a single GM. Since leachate through a single GM can be large if holes and flaws are involved, some type of CCL or GCL backup is often utilized. As a composite primary liner, the choices are GM/CCL or GM/GCL. For issues that will be discussed later, there are several compelling reasons to seriously consider the GM/GCL option.

Secondary Liner of a Double Liner System

Regulatory-wise, federal and most state agencies require a GM/CCL composite for the secondary liner. This is an excellent strategy when natural clay deposits are economically available. Since the CCL is placed directly on the soil or rock subgrade, large construction equipment can be used in its placement, and large compaction equipment can be used for its densification. Thus both low permeability and high shear strength can be attained without any danger of damage to underlying strata that do not contain geosynthetics. However, for sites where natural clay borrow materials are not available, consideration should be given to either a GM/GCL or GM/GCL/CCL composite.

Table 15.2   Leakage Flow Rates for Double-Lined Landfills

	No. of cells investigated	
Leakage detection layer flow rates	With CQA	Without CQA
Flow rate less than 5 gpad (50 lphd[a])	4	—
Flow rate in range of 5 to 20 gpad[b] (50 to 200 lphd)	4	1
Flow rate in range of 20 to 50 gpad (200 to 500 lphd)	3	—
Flow rate greater than 50 gpad (500 lphd)	—	4

[a] lphd = liter/ha-day.
[b] gpad = gal/acre-day.

Leak Detection Materials

Clearly, a granular soil (sand or gravel) can be used for leak detection in double liner systems, however, current practice generally utilizes a geonet. The reasons are as follows:

- savings in air space
- no perforated pipe system is necessary
- the danger of damage to underlying geosynthetics is avoided
- stability on side slopes can be achieved and
- construction is rapid and straightforward.

If the overlying material is a CCL or GCL, it is customary to thermally bond a geotextile to the geonet leak detection drain. The design of this geotextile is critical. If it is a woven, consideration must be given to the undesirable possibility of extrusion of overlying clay materials into the geonet. If it is a needle-punched nonwoven, consideration must be given to the possibility of excessive intrusion of the geotextile into the voids of the geonet. For these reasons, the suggested design must be laboratory tested for verification.

Leak Detection Flow Rates

While actual leakage rate measurements from double lined landfills is sparse, the information that is available speaks well for the concept. Table 15.2, after Bonaparte and Gross[8] clearly shows that 5 to 50 gal/acre-day (50 to 500 l/ha-day) is achievable with some type of construction quality assurance (CQA). The data also shows the clear necessity of having CQA so as to significantly reduce the leakage rates. The paper by Daniel and Koerner[9] in these proceedings describes CQA of landfills along with related activities.

Equally as important as the leak detection flow rates themselves is the identification of where the liquid is originating. Bonaparte and Gross[8] describe four possible scenarios;

- upper liner leakage of leachate;
- clay consolidation water from an overlying CCL;
- drainage layer compression water; and/or
- lower liner infiltration water.

Of these sources, clay consolidation water from a CCL can be enormous. For a 18 in. (450 mm) thick CCL placed at 3% wet of optimum water content, the initial flow rate can easily be in excess of 100 gal/acre-day (1000 l/ha-day). This amount will usually overshadow the liner leakage rate which is of primary interest. Using a GCL as the lower component of the primary liner avoids this troublesome situation.

LIQUIDS MANAGEMENT SYSTEMS

The flow, collection and removal of leachate from a landfill as shown in Figure 15.2 is called the liquids management system. Some important construction and design considerations are included in this section.

Leachate Quantities

With respect to leachate, both quality and quantity are of concern. Regarding *quality*, the leachate takes on the characteristics of the material being landfilled and is very clearly waste specific in this regard; its identification was described earlier. Regarding *quantity*, the leachate is both waste specific and site specific. In order to estimate leachate flow rates in a landfill a computer code under the acronym of HELP is available, Schroeder, et al.[10] This code considers the site hydrology, landfill geometry, type of solid waste, thickness of solid waste and other relevant parameters in a very sophisticated accounting of the liquid in the waste and in the local precipitation, to arrive at the predicted flow rate at the bottom of the landfill. The code is used by most designers and all regulators to quantify leachate flow rates.

Piping Systems

Within natural soil drainage systems there must be located a perforated pipe drainage system to collect and transmit the leachate to a down gradient sump. This piping system is usually HDPE pipe, of either smooth wall or profiled wall varieties. The perforations are usually holes for smooth wall and slots for profiled types. The design procedures follow standard pipe methods, i.e., pipe spacing via the mound model, pipe diameter via hydraulics theory, and pipe strength via the Iowa State formula, see Koerner.[2]

It is important to note that the use of a geonet drainage system avoids the use of pipes and is a distinct advantage for the use of geosynthetics.

Sumps

At the low end of the landfill or cell, the leachate enters a shallow depression in the liner system where it is collected for removal and proper handling or treatment. Note that with double lined systems, both the leachate collection and leak detection systems must have their own individual sumps. These shallow depressions are from 12 to 36 in. (300 to 900 mm) in depth and are very important insofar as their construction and inspection is concerned. It is in the sump areas that the hydraulic heads are the highest, and if leaks occur in the liner system they will result in relatively high flow rates out of the containment system.

Removal Manholes or Risers

Extending from the sump is either a vertical manhole or a sidewall riser. The vertical manhole option is being used less frequently due to downdrag forces occurring from the subsiding waste and landfill operation difficulties due to the physical obstruction of the manholes rising through the waste. Sidewall risers avoid these problems and are becoming widespread in their use. Sidewall risers will generally be of an adequate diameter so that a submersible pump can be placed inside and lowered from a shed down into the sump, e.g., 18 to 30 in. in diameter. The shed will also be used as the terminus of a smaller pipe used to monitor a leakage in a double lined system. This small leak detection pipe (typically 4 to 6 in. in diameter) will have a necessary geomembrane penetration of the primary liner at, or near, to the top of the slope.

Leachate Recycling

The Subtitle D regulations for nonhazardous landfills permit the landfill operator to recycle the extracted leachate back on top of the solid waste mass in what is known as *leachate recycling*. The reasons for this practice are to accelerate bioreactions within the waste and to extract heavy metals and organics on a periodic basis. Leachate recycling is ongoing at 20 to 30 landfills in the U.S. and is considered to be somewhat experimental at this time. It should be noted, however, that a form of leachate recycling has been practiced at various landfills for many years.

Daily Cover

Whatever the practice of liquids management, it is necessary that the leachate flow gravitationally through the waste into the leachate collection system above the primary liner. The daily soil cover placed over each lift of waste can have the tendency to block this desired downward flow path. If it does, the leachate remains perched in the waste and has even exited the landfill through its cover in the form of sidewall seeps. Neither of these features are desirable.

For the above reasons, coupled with the high cost of landfill air space and the cost of daily soil cover, a number of replacement materials are currently available. Collectively, they are called alternate daily cover materials, or ADCMs. Table 15.3 is modified from Pohland and Graven.[11]

Table 15.3 Types of Alternate Daily Cover Materials

- Polymer based materials (typically foams)
  - Rusmar™
  - Saniform™
  - Terrafoam™
  - Topcoat™

- Slurry spray materials (paper or wood chip based)
  - Con Cover™
  - Land-Cover™
  - Posishell™

- Sludges and indigenous materials (with or without binders)
  - Naturite/naturefill™
  - N-viro soil™
  - Chemfix™
  - Ash-based
  - Auto fluff
  - Foundry sand
  - Green waste/compost
  - Shredded tires

- Reusable geosynthetics (geotextiles or geomembranes)
  - Air space saver™
  - Aqua-shed™
  - Covertech™
  - Cormier™
  - FabriSoil™
  - Griffolyn™
  - Polyfelt®
  - Sanicover™
  - Typar®

Modified from Pohland and Graven.[11]

Covers or Closures

The cover, or closure, of a landfill is critically important since its performance must be assured over an extremely long lifetime. Lifetimes well beyond the 30-year post closure care period are often considered. Within the cover system are the following elements which must be evaluated and designed according to site specific and waste specific considerations.

- vegetative cover and top soil
- cover soil
- surface water drainage system (unless arid climate)
- composite barrier system (GM/CCL or GM/GCL)
- gas venting layer (required for municipal solid waste)
- final compacted cover soil over the solid waste mass

Details of these layers and their design elements are found in Daniel and Koerner[12] and are contained in these proceedings, see Landreth.[1]

CURRENT STATUS AND FUTURE CONSIDERATIONS

It appears as though the regulatory structure for nonhazardous (Subtitle D) and hazardous (Subtitle C) solid waste landfill liners is in place and, coupled with the leak detection rules, forms the basic strategy of the federal U.S. EPA.

There is movement within individual states, but clearly minimum technology guidance at the federal level has been set. The states obviously must meet, or exceed, these guidance regulations. The remaining issues, however, are felt to be on the immediate horizon.

Material Considerations

It should not be surprising that variations of the existing materials used for waste containment liner systems will come about. Hopefully, the newer materials will be environmentally superior and at a better benefit/cost ratio than existing materials. Clearly, there is robust activity in the development of new materials that will be offered to the waste containment industry. For example:

- engineered clays to attenuate specific containments,
- various coextruded geomembranes,
- new, or modified, types of geomembrane texturing,
- variations of existing GCLs,
- high compressive strength and flow rate geonets,
- various geomembrane protection materials, and
- enhanced long term durability and performance geosynthetics.

Design Issues

Design, being at the heart of a properly functioning waste containment system, is being actively pursued by many consultants and university researchers. Most significant in this regard are felt to be the following:

- side slope stability designs,
- seismic analysis and methods for limiting deformations,
- vertical and horizontal expansion techniques,
- leachate movement and treatment designs,
- liquids management designs for large areal landfills, and
- designs for individual component behavior at landfills of very large heights.

Material Testing Considerations

Both natural soils and geosynthetics require a bevy of test methods and establishment of testing protocols. ASTM is very active in this regard. Committee D18 on Soil/Rock handles the CCLs and natural drainage materials. Committee D35 on Geosynthetics handles the GCLs and other geosynthetics. Both of these ASTM committees are active, as well as standards institutes over the entire world. Indeed, for any type of design-by-function, one must counterpoint an allowable (or test) property with the appropriate required (or design) property. The tacit assumption is that one is dealing with engineering materials (natural soils and geosynthetics) and they must be analyzed and treated as such. Important test methods under current development are as follows:

- laboratory testing methods for CCLs
- chemical resistance of all types of geosynthetics
- endurance procedures for geosynthetics
- methods of fingerprinting geosynthetics
- new test methods for GCLs
- performance testing of all types of natural soil and geosynthetics
- field test pad evaluation for all types of natural soils and geosynthetics

Construction Considerations

Construction issues from both quality control and quality assurance perspectives are critically important, see Daniel and Koerner.[9,12,13] Significant new advances are as follows:

- new field testing and evaluation of CCLs,
- new automated seaming devices for geomembranes,
- new inspection methods for flaws in geomembranes, and
- enhanced critical path methods for sequential layer-by-layer construction.

Operations Considerations

The placement of solid waste at landfill sites can be of overriding significance not only from owner/operators perspective but also from the design and performance perspective. Indeed, failures of landfills have occurred and generally can be associated (at least in part) to operations at the site. Some new trends are felt to be as follows:

- utilization of earth pressures of the solid waste to stabilize side slopes,
- deep dynamic compaction to densify waste and gain air space,
- isolating and/or mixing waste so as to enhance leachate movement and avoid excessive clogging of leachate collection filters,
- baling of waste for optimization of placement and for stability concerns,
- covering waste to minimize, or even eliminate, leachate generation,
- enhanced biological degradation of waste to accelerate decomposition and gain air space, and
- mining of waste after adequate decomposition for material recovery and use of the residuals.

Field Performance Considerations

At this point in time there is a dire need to investigate and assess the field performance of solid waste landfills. The salient issues to be investigated are as follows:

- leak detection monitoring for double lined landfills,
- identification of the origin of the liquid in the leak detection layer,
- downgradient monitoring for all types of landfills,

- economic analysis of downgradient monitoring wells vs. double lined systems, and the associated risk assessment of doing with fewer (or no) monitoring wells,
- temperature and bioreactivity of the solid waste mass,
- leachate analysis as to the decomposition of the encapsulated solid waste mass,
- deformation of the cover systems (both total and differential settlement),
- behavior of landfills under seismic activity, and
- possible use of closed landfill facilities for recreational, aesthetic, or industrial use.

A U.S. EPA project on some of the above issues has recently been granted to a group consisting of Drexel University's Geosynthetic Research Institute, The University of Texas at Austin, and GeoSyntec Consultants in Atlanta, GA. Information will be forthcoming over the next three years.

ACKNOWLEDGMENTS

Financial support for this research was provided by the Geosynthetic Research Institute's consortium of member organizations and the U.S. EPA. The author is co-principal investigator (with D. E. Daniel of University of Texas at Austin and R. Bonaparte of GeoSyntec Consultants of Atlanta) on a U.S. Environmental Protection Agency project on the subject of waste containment liner systems. Financial support under cooperative agreement CR-821448 is sincerely appreciated. Mr. Robert E. Landreth is the past Project Officer and Mr. Daniel A. Carson is the current Project Officer.

REFERENCES

1. Landreth, R. E. (1997), Cover Systems for Waste Management Facilities, in *Municipal Solid Wastes: Problems and Solutions,* Landreth, R. E. and Rebers, P. A., Eds., CRC Press, Boca Raton, FL.
2. Koerner, R. M. (1994), *Designing with Geosynthetics*, 3rd Ed., Prentice-Hall Publ. Co., Englewood Cliffs, NJ.
3. Daniel, D. E. (1987), Earthen Liners for Land Disposal Facilities, *Proceedings, Geotechnical Practice for Waste Disposal*, Univ. of Michigan, ASCE, 21–39.
4. Goldman, L. J., Greenfield, L. I., Damle, A. S., Kingsburg, G. L., Northein, C. M., and Truesdale, R. S. (1988), *Design, Construction and Evaluation of Clay Liners for Waste Management Facilities*, EPA/530-SW-86-007-F.
5. Daniel, D. E. and Boardman, B. T. (1993), Report on *Workshop on Geosynthetic Clay Liners*, U.S. Environmental Protection Agency, EPA/600/R-93/171.
6. Harpur, W. A., Wilson-Fahmy, R. F., and Koerner, R. M., Evaluation of the Contact Between GCLs and GMs in Terms of Transmissivity, *Proceedings 7th GRI Conference*, IFAI Publ., St. Paul, MN, 138–149.
7. Narejo, D. B. (1994), Puncture Behavior of HDPE Geomembranes, Ph.D. Thesis, Drexel University, Philadelphia.

8. Bonaparte, R. and Gross, B. A. (1990), Field Behavior of Double Lined Systems, *Proceedings Waste Containment Systems*, ASCE, Geotech. Spec. Publ. #26, New York, 52–83.

9. Daniel, D. E. and Koerner, R. M. (1997), Construction Quality Assurance/Quality Control for Landfills, in *Municipal Solid Wastes: Problems and Solutions,* Landreth, R. E. and Rebers, P. A., Eds., CRC Press, Boca Raton, FL.

10. Schroeder, P. R., Morgan, J. M., Walsh, T. M., and Gibson, A. C. (1989), *Hydrologic Evaluation of Landfill Performance (HELP) Model,* Vols. 3-5, Reports of Interagency Agreement No. DW 21931425, U.S. EPA, Cincinnati, OH.

11. Pohland, F. and Graven, J. P. (1993), *The Use of Alternative Materials for Daily Cover at Municipal Solid Waste Landfills*, U.S. EPA Report EPA/600/R-93/172 (also NTIS PB93-227197).

12. Daniel, D. E. and Koerner, R. M. (1993a), Cover Systems, in *Geotechnical Practice for Waste Disposal*, Daniel, D. E., Ed., Chapman and Hall, New York, pp. 455–496.

13. Daniel, D. E. and Koerner, R. M. (1993b), *Quality Assurance and Quality Control for Waste Containment Facilities*, U.S. Environmental Protection Agency, EPA/600/R-93/182.

14. Giroud, J. P. and Bonaparte, R. (1989), Leakage Through Liners Constructed with Geomembranes, Part III: Composite Liners, *J. Geotextiles and Geomembranes*, 8, 2, 71.

Index

C

Milton Keynes UK
Ingram Content Group UK Ltd.
UKHW040445071024
449327UK00020B/1006